Environmental Damage to DNA and the Protective Effects of Phytochemicals

Environmental Damage to DNA and the Protective Effects of Phytochemicals

Bechan Sharma and Nitika Singh

CRC Press
Taylor & Francis Group
Boca Raton London New York

CRC Press is an imprint of the
Taylor & Francis Group, an **informa** business

First edition published 2021
by CRC Press
6000 Broken Sound Parkway NW, Suite 300, Boca Raton, FL 33487-2742

and by
CRC Press
2 Park Square, Milton Park, Abingdon, Oxon, OX14 4RN

Library of Congress Cataloging–in–Publication Data

Names: Sharma, Bechan, author.
Title: Environmental damage to DNA and the protective effects of phytochemicals / Bechan Sharma, Nitika Singh.
Description: First edition. | Boca Raton : CRC Press, [2022] | Includes bibliographical references and index.
Identifiers: LCCN 2021011489 | ISBN 9780367358228 (hardback) | ISBN 9781032055169 (paperback) | ISBN 9780429342059 (ebook)
Subjects: LCSH: Phytochemicals. | DNA damage--Environmental aspects. | Biochemical toxicology.
Classification: LCC QK865 .S525 2022 | DDC 572/.2--dc23
LC record available at https://lccn.loc.gov/2021011489

ISBN: 978-0-367-35822-8 (hbk)
ISBN: 978-1-032-05516-9 (pbk)
ISBN: 978-0-429-34205-9 (ebk)

Typeset in Times
by Deanta Global Publishing Services, Chennai, India

Contents

Preface

The book provides information on the toxicity of natural as well as synthetic chemicals in the living systems leading to DNA damage and on emergence of serious consequences or manifestations causing varied health hazards. In addition, the book may reflect on the possible applications of plants or plant extracts to protect the living cells from xenobiotics-mediated DNA damage.

This book caters to the need of varied audiences such as undergraduates, postgraduates, doctoral and post-doctoral researchers, scientists, teachers, scientific reporters, and industry or corporates interested in learning about molecular events involved in DNA damage by different xenobiotics and protection thereof by plant-based principles. The areas concerned with this title include Molecular Biology, Biotechnology, Environmental Sciences, Molecular Toxicology, Pharmacophore Design and Development, and Phytochemistry. It is intended to provide an informative introduction to this book and to give a balanced reasonably detailed account of the role of different phytochemicals isolated from toxic and non-toxic plant species in the DNA damage and/or protection.

Most scientific terms are defined and placed in context when they are earlier introduced or described. The chief aim of this particular book is to help readers to have current understanding as well as the concept and mechanisms involved. In order to achieve this, we have made the book illustrative using suitable figures, tables, and images. The textual presentation has been supplemented with examples and chemical reactions to make it more appealing. Connectivity of each topic with the general perception about the subject has been ensured by giving its perspective at the beginning of every chapter.

Acknowledgments

The authors express a deep sense of gratitude and thanks to their well-wishers, friends, and family members for providing continued affection, encouragement, and cooperation which inspired us to fulfill the task. We warmly acknowledge the work of the artists, typesetters, and other individuals including the contribution of the publication house and technical persons, who made the publication of this book possible. We are thankful to the honorable reviewers, who contributed their ideas, suggestions, and critiques on the manuscript. The authors express their gratitude to the University of Allahabad for providing an academic platform and the basic amenities such as the library, subscription, and Internet facilities. The financial assistance to one of the authors, NS, by University Grants Commission-New Delhi, in the form of a Research Fellowship is acknowledged. Gratitude, they say, cannot be seen, it can only be felt and feelings of heart are hard to describe. Above all, we remember the Almighty, who gave us the courage and strength, power and enthusiasm for this purpose.

B. Sharma, Ph.D.
Professor of Biochemistry (Cadre)
Department of Biochemistry
University of Allahabad
Allahabad 211002, UP
India
&
Nitika Singh, M.Sc.
Senior Researcher
Department of Biochemistry
University of Allahabad
Allahabad 211002, UP
India

Authors' biographies

Dr. Bechan Sharma is presently working as a Professor and Head, Department of Biochemistry Allahabad University, Allahabad. Prof. Sharma has earlier serviced as Assistant Professor at Sher-e-Kashmir University of Agricultural sciences and Technology, J&K, and as Associate Professor at Dr. RML Avadh University, Faizabad. Dr Sharma completed his B.Sc (Honours)1980, M.Sc. (Biochemistry)1982, and Ph.D. (Biochemistry)1988 from BHU-Varanasi. He qualified GATE in 1983 and CSIR (JRF-NET) in 1984 and then joined Central Drug Research Institute, Lucknow, for completing his doctoral research. The areas of his research interest include Molecular Biology of HIV/AIDS, Tropical Diseases (Filariasis/Malaria), and Biochemical Toxicology. He has received a number of Awards/Honors and successfully completed numerous important Academic/Administrative Assignments. He has 33 years of teaching/research experience. He has carried out several research projects and published over 230 research papers including book chapters, molecular methods in peer-reviewed International and National Journals/Books/Methods of repute with high impact factors. He has one US patent on HIV-1 genome structure based antiHIV-1 drugs design to his credit. He has supervised 18 PhD candidates and seven Post-Doctoral Research Fellows, and six are continuing research under his supervision. He is member/life member of several national/international scientific societies and has attended a number of symposia/conferences in India and abroad and presented his research papers therein. He is Chief-Editor/Associate Editor/Executive Editor and Member Editorial Board of 150 peer-reviewed International and National Journals. He has been acting as honorary reviewer for about 190 International/National scientific journals. Six of his books have been published by international publishers and two are in print. He has worked as a visiting scientist in USA for over three years and visited different labs at Italy, France, Iran, Thailand, Germany, Japan, Hongkong and Brazil to conduct different collaborative research projects related to HIV-1 molecular biology, molecular epidemiology, bioinformatics and HIV-1 genome structure based antiHIV-1drug design, etc. Prof. Sharma has recently been awarded ICMR's Senior Scientist Fellowship 2014–15 in Biomedical Sciences under Indo-USA joint research program. Prof. B. Sharma is a Fellow of Academy of Environmental Biology and Bioved Research Society.

Ms. Nitika Singh is currently pursuing D. Phil. research program at the Department of Biochemistry, Faculty of Science, University of Allahabad, India, on the research topic concerning xenobiotics-mediated biochemical perturbations in mammalian systems and its protection by plant-based principles. She has been a very meritorious student throughout her academic career and received several merit awards. After graduating from M.G. Kashi Vidyapeeth-Varanasi, she qualified in the National level examinations of University of Allahabad to pursue a Master's Program in Biochemistry and a Doctoral Research Program in Science during 2014 and 2016, respectively. She has also successfully completed a Diploma Course in French. She is an expert in handling several modern techniques applied in Biochemistry, Toxicology, and Molecular Biology research (Agarose gel electrophoresis, SDS and Native PAGE, Submarine gel electrophoresis, Centrifugation, DNA isolation, DNA damage and repair assay, and Enzymes assays using Spectrophotometer, Gel Doc Scanning, HPLC, Spectro-fluorometry, and Microscopy). She has published papers in many peer-reviewed international journals of repute. Ms Singh has attended several workshops, conferences, and seminars and presented her research papers/posters in many scientific meetings.

Introduction to the environmental factors and DNA damage

1

INTRODUCTION

Environmental factors can be explained as any factor, either abiotic or biotic, which impacts living systems. These may be responsible for causing non-hereditary diseases encouraged by exposure to a particular condition such as physical and mental abuse, diet, pathogens, toxins, radiation, and chemicals. The abiotic factors encompass physical agents, such as radiation (IR and UV/medical X-rays) and heat, and chemical agents, including pesticides, heavy metals, cigarette smoke, some food additives, chemotherapeutic drugs, solvents, airborne pollutants, and industrial chemicals. The biological factors comprise toxins derived from animal and plant sources, mainly their secondary metabolites including phytochemicals. Additionally, some viral and bacterial agents such as *Helicobacter pylori*, human immunodeficiency virus (HIV), and human papillomavirus have been revealed to possess the potential to damage DNA (Hadley, 2006; Fazzo et al., 2017; Singh et al., 2017; Singh and Sharma, 2019).

The exposome research involves the totality of human environmental exposures from conception onwards, which only deals with non-genetic exposures (Jones, 2016). DNA is constantly exposed to various factors involving both the endogenous processes (free radicals) and the external insults (physical and chemical factors), thereby posing a challenge to cell survival. Usually, the genomes of all living organisms, both plants and animals, are stable. Due to consistent exposure of the genome to numerous environmental factors, the DNA gets damaged, producing diverse hereditary syndromes, which get transferred from one generation to the next. Both natural and anthropogenic toxins enter into the body of an organism via several routes of exposure, for instance, dermal contact, ingestion, inhalation, injection, and accidental exposure. These chemicals are described to induce DNA impairment, cellular membrane damage, metabolic disorders, protein dysfunction, mutagenicity, cancer, and cell death (Singh and Sharma, 2019).

The environmental factors responsible for producing DNA damage are called genotoxic agents. These factors include radiations and loss of nucleotide base(s) through spontaneous hydrolysis of the glycosidic bond and attack by reactive agents such as reactive oxygen and nitrogen species and alkylating agents. One of the most abundant lesions in DNA is the abasic or apurinic/apyrimidinic (AP) site. This lesion is mutagenic and can block DNA replication and transcription, leading to cell death. DNA bases also can become oxidized and one of the prominent oxidized bases is 8-oxo-7,8-dihydro-2'-deoxyguanosine (8-oxoG) in DNA. Furthermore, the oxidized guanine base can be formed in the dNTP pool (8-oxodGTP), and the nucleotide pool can contain enough 8-oxodGTP to promote mutagenesis (Çağlayan and Wilson, 2015). Also, these genotoxic agents induce their effects by distorting the DNA structure by breakage of hydrogen bonds responsible for stability of DNA strands. The types of DNA damage include oxidative damage, hydrolytic damage, bulky adduct formation, mismatch of bases, alkylation of bases, and DNA strand breaks (Klarer and McGregor, 2011; Tuteja et al., 2001). This chapter focuses on the sources, types of environmental DNA-damaging agents, and their impacts on DNA. In addition, types of DNA damage have been discussed.

SOURCES OF ENVIRONMENTAL DNA-DAMAGING AGENTS

Organisms have evolved to efficiently respond to DNA insults that result from either endogenous sources (cellular metabolic processes) or exogenous sources (environmental factors). Sources of endogenous insults of DNA damage include oxidation, hydrolysis, and alkylation reactions, as well as mismatch of DNA bases. The sources of exogenous DNA damage include radiations (IR and UV) and various chemical mediators (Hakem, 2008). Life has evolved in a world containing significant levels of IR. The radiations have always been present around us. Humans are exposed to them through medical treatments and activities involving radioactive materials. Organisms are persistently exposed to small amounts of IR from the environment as a part of their daily activities, known as background radiation. Commonly, cosmic radiation (CR) comprises fast-moving particles which occur in space and also are radiated from many other sources such as the Sun and celestial events in the universe. Mostly, cosmic rays are protons but they can be some other particles or any wave energy. Some of this IR penetrate the earth's atmosphere and is absorbed by humans/animals, the phenomenon of which is known as natural radiation exposure. A small amount of radioactive minerals is naturally present in food and drinking water, as typically vegetables are grown in soil using groundwater that contains moderate amounts of radioactive minerals. Upon ingestion of these items by humans or animals, exposure to natural radiation takes place (Aemr et al., 2012).

Radiations have numerous perspectives in medicine. For instance, X-ray machines use their radiation for detecting broken bones and also for the diagnosis of diseases. Additionally, nuclear medicine uses radioactive isotopes for the diagnosis and treatment of diseases including cancer (WHO, 2016). Radiations have several applications in industrial processes including nuclear gauges (used to build roads) and density gauges (used to measure the flow of material through pipes in factories). Radiations are also used in detection of smoke, in some glow-in-the-dark exit signs, for estimating reserves in oil fields, and for sterilization, which is carried out using large, heavily shielded irradiators. During fire incidents, large thermal stress responsible for a rapid increase in temperature within a short time may originate, which may induce serious damage to DNA i.e. A-DNA, B-DNA and Z-DNA. Solar radiation causes temperature changes.

There are several chemical toxicants that are released into the atmosphere from natural as well as anthropogenic sources. The sources of chemical toxicants in our homes include kitchens, bathrooms, laundries, and sanitary facilities. The emission of waste material from numerous activities including mining and manufacturing of batteries, and smoke, fume, dust particles of chemicals are other sources of chemical toxicants. The sources of heavy metals and pesticides include agricultural practices, working in pesticide and heavy metal infested environments, and household practices (Singh et al., 2017). There are numerous food contaminants responsible for damaging DNA that are introduced by human beings and are also naturally occurring in air, water, and soil. Food processing and food packaging materials may also lead to chemical contamination due to the migration of some harmful substances into foods (Rather et al., 2017).

The sources of biological agents include animals, bacteria, viruses, and plants. These are mainly secondary metabolites of animals and plants. Two main groups of biological agents are categorized as occupational biohazards: (1) allergenic and/or toxic agents forming bioaerosols, causing occupational diseases of the respiratory tract and skin, primarily in agricultural workers; and (2) zoonotic agents causing zoonoses and other infectious diseases that could be spread by ticks or insect vectors, through various exposure routes. Bioaerosols are biological particles of organic dust and/or droplets suspended in air, such as viruses, bacteria, endotoxin, fungi, secondary metabolites of fungi, particles of feces, bodies of mites and insects, and feather, hair, feces, and urine of birds and mammals (Dutkiewicz et al., 2011; Rim and Lim, 2014).

TYPES OF ENVIRONMENTAL DNA-DAMAGING AGENTS

The environmental DNA-damaging agents are placed under three different categories: (1) physical DNA-damaging agents, (2) chemical DNA-damaging agents, and (3) biological DNA-damaging agents. The causes of DNA lesions are mentioned in Table 1.1.

(1) Physical DNA-damaging agents

Physical agents such as UV and IR radiations have been reported to produce reactive oxygen species, which tend to damage cellular systems, including DNA. The UV radiation has an electromagnetic spectrum spanning from 100 nm to 400 nm, which is further subdivided into UV-A (400320 nm), UV-B (320290 nm), and UV-C (290100 nm). UV-C radiations are usually blocked by the atmospheric ozone layer, whereas UV-A and UV-B are mainly responsible for serious DNA damage. In addition, exposure to high temperature or heat also causes DNA damage (Gill et al., 2015; Todorova, 2019).

TABLE 1.1   Types and Causes of DNA Lesions

Type of DNA Lesion	Examples/Causes of DNA Lesion
Alterations in bases	Due to ionizing radiation and alkylating agents such as ethyl-methane sulfonate
Bulge due to insertion/ deletion of nucleotide bases	Due to intercalating agents such as acridines, caused by addition or deletion of nucleotide during replication or recombination
Cross-linked strand breaks	Due to covalent linkage of two strands by bifunctional alkylating agents including mitomycin C
3′-deoxyribose segmentation	Free radical-induced disruption of deoxyribose structure via strand breaks
Incorrect base	Mutation affecting 3′-5′ exonuclease proofreading of incorrectly incorporated bases
Linked pyrimidines	UV irradiation leads to the formation of cyclobutyl dimers (usually thymine dimers)
Single and/or double-strand breaks	Due to breakage of phosphodiester bond by ionizing radiation or chemical agents such as bleomycin

(2) Chemical DNA-damaging agents

The chemical agents causing DNA damage comprise various heavy metals responsible for damaging DNA by generating double-strand breaks (DSBs). The biochemical functions of certain key proteins are inhibited by chemicals; these proteins play an essential role in different DNA repair pathways. However, the indiscriminate use of pesticides in agricultural and health practices has also been observed to cause DNA damage, mutation, and chromosomal alterations. Copper is present in the nucleus of cells and is associated with chromosomes and DNA bases. The activation of DSBs of DNA can be initiated by copper in combination with some pesticides such as fluoxastrobin and imazamox. Industrial chemicals including brominated flame retardants, fluoroalkyl substances, dichloroethane (an industrial solvent), butadiene released from rubber factories, vinyl chloride, and hydrogen chloride from plastic manufacturing plants are investigated to have serious DNA-damaging potential since they interfere with the repair machinery involved in fixing DNA aberrations. Food additives including coloring agents such as red no. 40 (Allura red AC), Yellow no. 5 (Tetrazine), and Yellow no. 6 (sunset yellow) have been grouped under chemical agents with carcinogenic properties. Also, some food preservatives like nitrate and nitrite (highly toxic and carcinogenic agents) have been reported to treat bacterial growth. Additionally, some cross-linking agents like mitomycin C and cisplatin, alkylating agents, aromatic compounds, and fungal and bacterial toxins are also reported to have DNA-damaging potential (Morales, 2016; Pellacani et al., 2012).

(3) Biological DNA-damaging agents

Biological factors synthesized by both plant and animal species include metabolic byproducts and free radicals [ROS or reactive nitrogen species (RNS)]. There are some plant secondary metabolites such as nicotine, the major alkaloid present in tobacco, that are reported to induce carcinogenesis. Nicotine has been shown to have the potential for tumor promotion by causing DNA damage in many human epithelial and non-epithelial cells. Sanguinarine (an alkaloid) extracted from the plant *Argemone mexicana* has been reported to cause chromosomal aberrations, micronucleus formation, and DNA damage using comet assay in mice. Sanguinarine increases keratin formation and tumor promotion by inhibiting the activity of epidermal histidase (Ansari et al., 2004; Ghosh and Mukherjee, 2016).

TYPES OF DNA DAMAGE

The utmost important consequence of oxidative stress in the body is believed to cause damage to DNA. The modifications in DNA can occur in several ways that can lead to mutations and genomic instability. Finally, it could develop into various types of cancers such as those of breast, colon, and prostate. In this section we will discuss various types of DNA damage including oxidative damage, hydrolytic damage, DNA strand breaks, and others. Sources of DNA damage can be roughly divided into two main groups: endogenous damage and exogenous damage. Endogenous damages refer to damage caused by ROS or some agents produced from normal metabolic byproducts, while exogenous damages are caused by external agents, like radiations, including UV and X-rays; plant toxins; chemicals; viruses; and so on (Hakem, 2008; Broustas and Lieberman, 2014) (Figure 1.1).

OXIDATION OF BASES

Oxidative damage of DNA induced by several chemical and physical factors appears to be associated with cancer. Oxidative DNA damage is an inevitable consequence of cellular metabolism, with a propensity for increased levels of toxic insult. It can result in damage to all four bases and the deoxyribose. One of the most abundant and most easily measured products of this oxidation is

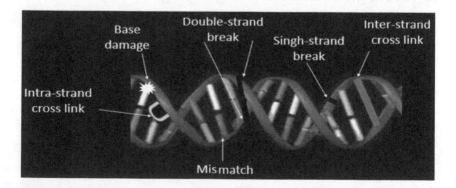

FIGURE 1.1.   The possible types of damage to genomic DNA. There are five main types of damage to DNA due to endogenous cellular processes.

8-oxo-7,8-dihydro-2'deoxyguanine (8-oxo-dG). Oxidative DNA damage refers to the oxidation of specific bases. 8-hydroxydeoxyguanosine (8-OHdG) is the most common marker for oxidative DNA damage and can be measured in virtually any species (Cooke et al., 2003). It is formed and enhanced most often by chemical carcinogens. Similar oxidative damage can occur in RNA with the formation of 8-OHG (8-hydroxyguanosine), which has been implicated in various neurological disorders. This type of damage to DNA involves the generation of DNA strand interruptions from reactive oxygen species. The peroxyl radicals (ROO.) are important oxidants found in cells are able to react with the DNA bases (Simandan et al., 1998).

HYDROLYSIS OF BASES

The covalent structure of DNA is unstable in aqueous solutions. It tends to hydrolyze to its monomeric components, which are themselves subject to various hydrolytic reactions. These processes are slow when compared to most familiar chemical reactions. Hydrolytic DNA damage involves deamination, depurination, and depyrimidination, or the total removal of individual bases. Loss of DNA bases, known as apurinic/apyrimidinic (AP) sites, can be particularly mutagenic and if left unrepaired can inhibit transcription. Hydrolytic damage may result from the biochemical reactions of various metabolites as well as the overabundance of reactive oxygen species (Shapiro, 1981; Williams, 2004).

ALKYLATION OF BASES

DNA alkylation refers to the addition of alkyl groups to specific bases, resulting in alkylation products such as O2-alkylthymine, O4-alkylthymine, O6-methylguanine, and O6-ethylguanine, which cause DNA mutations. In the case of methylation, it involves formation of 7-methylguanine, 1-methyladenine, 6-O-Methylguanine (Grady et al., 2007). Alkylating drugs are the oldest class of anticancer drugs still commonly used; they play an important role in the treatment of several types of cancers. Most alkylating drugs are monofunctional methylating agents (e.g., temozolomide [TMZ], -methylnitronitrosoguanidine [MNNG], and dacarbazine), bifunctional alkylating agents such as nitrogen mustards (e.g., chlorambucil and cyclophosphamide) or chloroethylating agents (e.g., nimustine [ACNU], carmustine [BCNU], lomustine [CCNU], and fotemustine) (Kondo et al., 2010).

MISMATCH OF BASES

Mismatch of bases, due to errors in DNA replication, leads to a situation where the wrong DNA base is stitched into place in a newly forming DNA strand, or a DNA base is skipped over or mistakenly inserted. DNA mismatch (MM) is a DNA defect occurring (1) when two non-complementary bases are aligned in the same base-pair step of a duplex DNA; (2) during replication of DNA; (3) during heteroduplex formation; and (4) due to mutagenic chemicals, ionizing radiation, or spontaneous deamination. While MMs are well tolerated in RNA, they are quickly corrected in DNA by the mismatch repair (MMR) proteins.

An MM formed by non-complementary purine (Pur)·pyrimidine (Pyr) bases is defined as 'transduction', and in the case of Pur·Pur or Pyr·Pyr pairs, 'transversion'. MMs introduce major changes in the canonical (Watson-Crick) recognition rules and are expected to produce major alterations in the structure and stability of the DNA double helix, especially in the proximity of the MM site (Rossetti et al., 2015).

BULKY ADDUCT FORMATION

The bulky DNA adducts, which usually have complex structures, are particularly important because of their biological relevance. The so-called bulky DNA adducts are formed by the covalent binding of those chemical carcinogens with large size, such as PAHs and aromatic amines, to various sites on DNA bases. These adducts also include exocyclic DNA bases such as the etheno, propano, and benzetheno adducts formed by respective bifunctional compounds. These bulky adducts represent a major and important class of DNA damage originating from exposure to cigarette smoke. One characteristic of these bulky adducts is that they tend to significantly disrupt the DNA helical structure and block Watson-Crick base pairing. They are usually highly mutagenic, as exemplified by the PAH-DNA adducts and exocyclic DNA adducts (Hang, 2010).

Different exogenous agents cause different damages. UV damage, alkylation/methylation, X-ray damage, and oxidative damage are examples of induced damage. Spontaneous damage can include the loss of a base, deamination, sugar ring puckering, and tautomeric shift. For instance, UV-B light causes cross-linking between adjacent cytosine and thymine bases, creating pyrimidine dimers. This is called direct DNA damage. UV-A light mostly creates free radicals. The damage caused by free radicals is called indirect DNA damage. Ionizing radiation such as that created by radioactive decay or in cosmic rays causes breaks in DNA strands. Low-level ionizing radiation may induce irreparable DNA damage (leading to replicational and transcriptional errors needed for neoplasia or may trigger viral interactions), leading to premature aging and cancer (Acharya, 1975; Acharya, 1976; Acharya, 1977). Thermal disruption at elevated temperature increases the rate of depurination (loss of purine bases from the DNA backbone) and single-strand breaks. For example, hydrolytic depurination is seen in the thermophilic bacteria, which grow in hot springs at 40–80 °C (Madigan and Martino, 2006; Ohta et al., 2006). The rate of depurination (300 purine residues per genome per generation) is too high in these species to be repaired by normal repair machinery; hence a possibility of an adaptive response cannot be ruled out. Industrial chemicals such as vinyl chloride and hydrogen peroxide, and environmental chemicals such as polycyclic aromatic hydrocarbons found in smoke, soot, and tar create a huge diversity of DNA adducts-ethenobases, oxidized bases, alkylated phosphotriesters, and cross-linking of DNA just to name a few.

CONCLUSION

The cellular DNA responsible for regulating each of the genetic characteristics of a living system is always prone to damage by several environmental factors comprising physical, chemical, and biological agents. The physical factors mainly include various radiations of varying energy levels and heat. The chemical agents mainly involve anthropogenic or synthetic chemicals. The biological agents primarily consist of bacterial and virus species which possess abilities to modulate DNA and its functionality within the cells. In addition, the generation of excess free radicals due to any stress and ingestion or exposure to anthropogenic chemicals or pathogenic microbes indirectly cause DNA damage due to generation of oxidative stress, resulting in varied inheritable genetic defects. The types of DNA damage caused due to exposure to different environmental factors include alterations in bases, bulge due to insertion/deletion of nucleotide bases, cross-linked strand breaks, 3′-deoxyribose segmentation, incorrect base, linked pyrimidines, and single and/or double-strand breaks.

REFERENCES

Acharya, P.V.N. 1975. The effect of ionizing radiation on the formation of age-correlated oligo Deoxyribo nucleo Phospheryl peptides in mammalian cells; 10th International Congress of Gerontology, Jerusalem; Abstract No. 1. Work done while employed by Dept. of Pathology, University of Wisconsin, Madison.

Acharya, P.V.N. 1976. Implications of the action of low-level ionizing radiation on the inducement of irreparable DNA damage leading to mammalian aging and chemical carcinogenesis; 10th International Congress of Biochemistry, Hamburg, Germany, University of Wisconsin, Madison; Abstract No. 01-1-079. Work done while employed by Dept. of Pathology.

Acharya, P.V.N. 1977. Irreparable DNA-damage by industrial pollutants in pre-mature aging, chemical carcinogenesis and cardiac hypertrophy: Experiments and theory; 1st International Meeting of Heads of Clinical Biochemistry Laboratories, Jerusalem, Israel. April 1977. Work Conducted at Industrial Safety Institute and Behavioral Cybernetics Laboratory, University of Wisconsin, Madison.

Afify, A.E.M.R., El-Beltagi, H.S., Aly, A.A., El-Ansary, A.E. 2012. The impact of γ-irradiation, essential oils and iodine on biochemical components and metabolism of potato tubers during storage. *Notulae Botanicae Horti Agrobotanici Cluj-Napoca* 40(2); 129–139.

Ansari, K.M., Chauhan, L.K., Dhawan, A., Khanna, S.K., Das, M. 2004. Unequivocal evidence of genotoxic potential of argemone oil in mice. *International Journal of Cancer* 112(5); 890–895.

Broustas, C.G., Lieberman, H.B. 2014. DNA damage response genes and the development of cancer metastasis. *Radiation Research* 181(2); 111–130.

Çağlayan, M., Wilson, S.H. 2015. Oxidant and environmental toxicant-induced effects compromise DNA ligation during base excision DNA repair. *DNA Repair (Amst)* 36; 86–90.

Cooke, M.S., Evans, M.D., Dizdaroglu, M., Lunec, J. 2003. Oxidative DNA damage: Mechanisms, mutation, and disease. *The FASEB Journal* 17(10); 1195–1214.

Dutkiewicz, J., Cisak, E., Sroka, J., Wójcik-Fatla, A., Zając, V. 2011. Biological agents as occupational hazards - selected issues. *Annals of Agricultural and Environmental Medicine* 18(2); 286–293.

Fazzo, L., Minichilli, F., Santoro, M., Ceccarini, A., Della, S.M., Bianchi, F., Comba, P., Martuzzi, M. 2017. Hazardous waste and health impact: A systematic review of the scientific literature. *Environmental Health* 16; 1–11.

Ghosh, I., Mukherjee, A. 2016. Argemone oil induces genotoxicity in mice. *Drug and Chemical Toxicology* 39(4); 407–411.

Gill, S.S., Anjum, N.A., Gill, R., Jha, M., Tuteja, N. 2015. DNA damage and repair in plants under ultraviolet and ionizing radiations. *Scientific World Journal* 2015; 250158.

Grady, W.M., Ulrich, C.M. 2007. DNA alkylation and DNA methylation: Cooperating mechanisms driving the formation of colorectal adenomas and adenocarcinomas? *Gut* 56(3); 318–320.

Hadley, C. 2006. The infection connection: Helicobacter pylori is more than just the cause of gastric ulcers-it offers an unprecedented opportunity to study changes in human microecology and the nature of chronic disease. *EMBO Reports* 7(5); 470–473.

Hakem, R. 2008. DNA-damage repair; the good, the bad, and the ugly. *The EMBO Journal* 27(4); 589–605.

Hang, B. 2010. Formation and repair of tobacco carcinogen-derived bulky DNA adducts. *Journal of Nucleic Acids* 2010; 709521.

Jones, D.P. 2016. Sequencing the exposome: A call to action. *Toxicology Reports* 3; 29–45.

Klarer, A.C., McGregor, W. 2011. Replication of damaged genomes. *Critical Reviews in Eukaryotic Gene Expression* 21(4); 323–336.

Kondo, N., Takahashi, A., Ono, K., Ohnishi, T. 2010. DNA damage induced by alkylating agents and repair pathways. *Journal of Nucleic Acids* 2010; 1–7.

Madigan, M.T., Martino, J.M. 2006. *Brock Biology of Microorganisms* (11th ed.). Pearson, p. 136.

Morales, M.E., Derbes, R.S., Ade, C.M., Ortego, J.C., Stark, J., Deininger, P.L., Roy-Engel, A.M. 2016. Heavy metal exposure influences double strand break DNA repair outcomes. *PLOS ONE* 11(3); e0151367.

Pellacani, C., Buschini, A., Galati, S., Mussi, F., Franzoni, S., Costa, L.G. 2012. Evaluation of DNA damage induced by 2 polybrominated diphenyl ether flame retardants (BDE-47 and BDE-209) in SK-N-MC .Cells. *International Journal of Toxicology* 31(4); 372–379.

Rather, I.A., Koh, W.Y., Paek, W.K., Lim, J. 2017. The sources of chemical contaminants in food and their health implications. *Frontiers in Pharmacology* 8; 1–8.

Rim, K.T., Lim, C.H. 2014. Biologically hazardous agents at work and efforts to protect workers' health: A review of recent reports. *Safety and Health at Work* 5(2); 43–52.

Rossetti, G., Dans, P.D., Gomez-Pinto, I., Ivani, I., Gonzalez, C., Orozco, M. 2015. The structural impact of DNA mismatches. *Nucleic Acids Research* 43(8); 4309–4321.

Shapiro, R. 1981. *Damage to DNA Caused by Hydrolysis.* doi:10.1007/978-1-4684-7956-0_1.

Simandan, T., Sun, J., Dix, T.A. 1998. Oxidation of DNA bases, deoxyribonucleosides and homopolymers by peroxyl radicals. *Biochemical Journal* 335(2); 233–240.

Singh, N., Gupta, V.K., Kumar, A., Sharma, B. 2017. Synergistic effects of heavy metals and pesticides in living systems. *Frontiers in Chemistry* 5; 1–9.

Singh, N., Sharma, B. 2019. Role of toxicants in oxidative stress mediated DNA damage and protection by phytochemicals. *EC Pharmacology and Toxicology* 7(5); 325–330.

Todorova, P.K., Fletcher-Sananikone, E., Mukherjee, B., Kollipara, R., Vemireddy, V., Xie, X.J., Guida, P.M., Story, M.D., Hatanpaa, K., Habib, A.A., Kittler, R., Bachoo, R., Hromas, R., Floyd, J.R., Burma, S. 2019. Radiation-induced DNA damage cooperates with heterozygosity of TP53 and PTEN to generate high grade gliomas. *Cancer Research* 14. pii:canres.0680.2019.

Toshihiro, O., Tokishita, S., Kayo, M., Jun, K., Masahide, S., Hideo, Y. 2006. UV sensitivity and mutagenesis of the extremely thermophilic Eubacterium Thermus thermophilus HB27. *Genes and Environment* 28(2); 56–61.

Tuteja, N., Singh, M.B., Misra, M.K., Bhalla, P.L., Tuteja, R. 2001. Molecular mechanisms of DNA damage and repair: Progress in plants. *Critical Reviews in Biochemistry and Molecular Biology* 36(4); 337–397.

WHO. 2016. Ionizing radiation, health effects and protective measures, https://www.who.int/news-room/fact-sheets/detail/ionizing-radiation-health-effects-and-protective-measures.

Williams, N.H. 2004. DNA hydrolysis: Mechanism and reactivity. In: *Artificial Nucleases. Nucleic Acids and Molecular Biology*, vol. 13, Zenkova, M.A. (eds.), Springer, Berlin, pp. 3–17.

Molecular mechanisms of physical environmental factors as DNA-damage agents with special reference to (A) Radiations and (B) Temperature

2

INTRODUCTION

Physical environmental factors known to play vital roles in the etiology of human cancer include physical carcinogens such as UV, IR, and X-ray irradiation (Minamoto et al., 1999). One of the vital goals of environmental monitoring is to detect potentially hazardous elements. The presence of any genotoxin within any organism does not represent a risk of hazard necessarily but when its concentration reaches beyond a critical level and when it binds to biologically important macromolecules such as DNA, protein, and lipid, it initiates a potentially damaging cascade. Further, all the organisms are exposed to reactive oxygen species (ROS) on a day-to-day basis that are capable of damaging DNA. Proper functioning of all living organisms depends on faithful maintenance of genomic information. Even though it has been observed that the information stored is stable and safe, the integrity of DNA is continuously challenged by numerous genotoxic assaults as well as environmental stress. Some of the essential physiological functions including oxidative respiration and lipid peroxidation generate ROS that are responsible for damaging DNA.

13

Moreover, environmental physical factors including UV, IR, and heat present a number of combustion products in the air, which are capable to induce a wide variety of DNA lesions. It has been estimated that in an average mammalian cell ten to a hundred thousand DNA lesions are introduced each day. Additionally, spontaneous hydrolysis of nucleotide bases induces non-instructive abasic (AP) sites. The consequences of DNA damage are diverse and adverse. For instance, acute cellular effects arise from altered gene transcription and DNA replication resulting in abnormal cellular responses, senescence/ apoptosis. The lesions of the DNA affect the proper segregation of chromosome during cell division and finally result in chromosomal aberrations. The errors because of DNA damage during the replication may result in irreversible mutations. Chromosomal aberrations and mutations in coding genes may also lead to production of abnormal proteins and finally emergence of various diseases including cancer (Dinant et al., 2008).

Radiation-induced DNA damage is a key event which considerably affects the normal life processes of all organisms. DNA is one of the most important macromolecules, and the stability of DNA is very important for proper functioning and maintenance of all living systems. UV radiation, a very powerful agent, exerts adverse effects on the genome stability. Ultraviolet and other types of radiations can damage DNA by introducing DNA strand breaks. This involves a cut in one or both DNA strands. The DNA double-strand breaks are especially dangerous and can be mutagenic, since they can potentially affect the expression of multiple genes. UV-induced damage can also result in the production of pyrimidine dimers, where covalent cross-links occur in cytosine and thymine residues. The adverse effects include the carcinogenic nature of radiation that can modify the normal state of life by introducing varied types of mutagenic and cytotoxic lesions to the DNA including cyclobutane-6-4 photoproducts (6-4PPs), pyrimidine dimers (CPDs), their Dewar valence isomers (Dewar valence isomers, the third type of environmentally relevant DNA photoproducts induced by solar radiation), and DNA strand breaks by interfering the genome integrity and may also affect the normal life processes of all living systems ranging from prokaryotic to mammalian systems. CPD and 6-4PP are the most frequent DNA mutations found in the p53 protein in skin cancers. Pyrimidine dimers can disrupt polymerases and prevent proper replication of DNA (Rastogi et al., 2010). The groups of UV radiation include UV-A: 315–400 nm, UVB: 280–315 nm, and UV-C: <280 nm. Among these, UV-A does not have significant efficacy of inducing DNA damage, as UV-A is not absorbed by native DNA. It has been observed that UV-A as well as visible light energy up to 670–700 nm wavelengths are able to produce singlet oxygen ($^1O_2$) which can be responsible for damaging DNA by indirect photosensitization reactions. The UV-B radiation introduces contrary effects on diverse habitats; however, most of the extra-terrestrial UV-B radiation is absorbed by the stratospheric ozone. The UV-C radiation is quantitatively absorbed by oxygen and ozone in the earth's atmosphere; hence it does not show much harmful effects on biota. Solar UV radiation is responsible for a wide range of biological effects including alteration in the structure of DNA. Although, UV-B radiation has less than 1% of total solar energy, it is a highly active component of the solar radiation that brings about chemical modification in DNA and changes its molecular structure by the formation of dimers (Rastogi et al., 2010).

Heat stress including heat shock and hyperthermia is one of the most studied as well as well-defined complex stress factors. Heat stress to the cell involves most of the sub-cellular compartments as well as metabolic processes (Kantidze et al., 2016). It has been reported that exposure of cells to heat stress shows an enhanced sensitivity for agents which are responsible for inducing double-stranded DNA breaks (DSBs), especially the ionizing radiation. This particular phenomenon is known as heat-radio sensitization. It has also been observed that heat stress can inhibit the DNA repair system (Iliakis et al., 2008). Indeed, several studies have revealed that heat stress can inhibit the key components of virtually all repair systems. Although heat-stress response has been extensively studied for decades, very little is known about its effect on nucleic acids and nucleic acid-associated processes. Heat stress is not only responsible for inhibition of DNA repair but also damage to DNA (Figure 2.1).

It has been reported that heat stress can lead to the accumulation of deaminated cytosine, 8-oxoguanine, and apurinic DNA sites (AP-sites) in a cell. It can be suggested that such DNA damage, as well as single-stranded DNA breaks (SSBs), is passively accumulated in the cell due to heat stress-induced inhibition of excision repair systems. A more interesting and controversial question is related to the nature of heat stress-induced DSBs, as well as the possibility of active heat stress induction of SSBs. For a long time, it was believed that heat stress does not induce DSBs, but rather leads to the generation of SSBs, which are formed as a result of inhibition of DNA replication due to hyperthermia (Kantidze et al., 2016). Chapter 2 illustrates the underlying mechanisms of actions of physical environmental factors involved in DNA damage.

MECHANISMS OF ACTION

The mechanisms of oxidative DNA damage have not been elucidated properly. Free radicals, commonly known as reactive oxygen species, contain single unpaired electron in their outer most orbit. The excessive production of free radicals results in the depletion of antioxidants *in vivo* and causes an imbalance

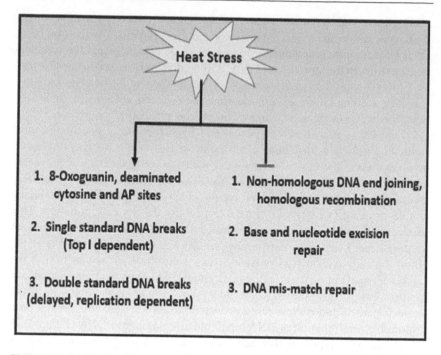

FIGURE 2.1.   Role of heat stress in induction of DNA damage and inhibition of repair pathways.

between the levels of free radicals and the antioxidant defense system of the body, which finally leads to generation of oxidative stress mediated damage. The production of 8-hydroxydeoxyguanosine (8-OHdG) is the most common biomarker of oxidative DNA damage. The free radicals have also been shown to remove an individual base generating abasic (apurinic / apyrimidinic; AP) sites into DNA. Other mechanism involved in DNA damage depends on the source of DNA damage. For example, reactive oxygen species (ROS) and reactive nitrogen species (RNS) induce oxidative base modifications as discussed above, infra-red typically leads to single (SSB)- and double-strand breaks (DSB), respectively; DNA alkylation leads to adduct and interstrand cross-link (ICL) formation, and UV radiation triggers the formation of thymidine dimers. The primary substances responsible for absorbing UV radiation include molecular oxygen, oxides of sulfur and nitrogen, and organic compounds such as aldehydes. A molecule of oxygen is split into reactive oxygen species after the absorption of UV radiation below 240 nm.

DNA damage results in (i) misincorporation of bases during replication process, (ii) hydrolytic damage, which results in deamination of bases, depurination, and depyrimidination, (iii) oxidative damage caused by direct interaction of ionizing radiations with the DNA molecules, as well as mediated by UV radiation-induced free radicals or reactive oxygen species and (iv) alkylating agents that may result in modified bases. The hydrolytic deamination (loss of an amino group) can directly convert one base to another; for example, deamination of cytosine results in uracil and at much lower frequency adenine to hypoxanthine (Rastogi et al., 2010).

The exposure of radiations of UV and ionizing radiations and certain genotoxic chemicals may result in single as well as double DNA strand breaks. Among different types of damages, DNA double-strand breaks (DSBs) are the most deleterious, since they affect both strands of DNA and can lead to the loss of genetic material. At high concentrations, the oxygen-free radicals or the reactive oxygen species (ROS) can induce damage to cellular structure and its constituents such as lipids, proteins as well as DNA and result in oxidative stress, which has been implicated in a number of human diseases. The hydroxyl radicals (OH∘) can damage all components of DNA molecules such as purine and pyrimidine bases and also the deoxyribose backbone, inhibiting the normal functions of the cell (Halliwell and Gutteridge, 2007; Valko et al., 2007).

DOUBLE-STRAND BREAKS (DSBS) OF DNA MEDIATED BY UV-RADIATION

It has been known for a long time that UV-irradiated cells undergoing replication of DNA involves generation of DSBs. Most of the DSBs in cells have been recorded under UV-B irradiation (Rastogi et al., 2010). DNA lesions including CPDs and 6-4PPs may induce primary and secondary breaks to the DNA, respectively. Generally, these lesions are correlated with the transcription as well as replication blockages that may lead to production of DSBs at the sites of collapsed replication forks of CPDs-containing DNA (Limoli et al., 2002; Batista et al., 2009). Similarly, Dunkern and Kaina (2002) have also observed UVC-mediated DNA DSBs arising from replication of damaged DNA. The UV-induced replication arrest in the xeroderma pigmentosum variant (XPV) followed by the accumulation of Mre11/Rad50/Nbs1 complex and phosphorylated histone H2AX (γ H2AX) in large nuclear foci at sites of stalled replication forks suggests that the UVR-mediated damage leads to the formation of DSBs during the course of replication arrest (Takahashi and Ohnishi, 2005). A number of pathways have been considered for the formation of DSBs at a stalled replication fork. Furthermore, replication stresses may trap topoisomerase I (Top1) cleavage complexes to generation of DSBs by preventing

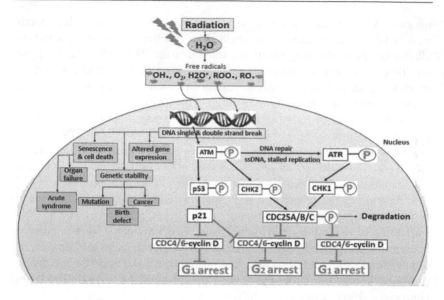

FIGURE 2.2. Involment of radiations in generation of free radicals causing DNA damage.

Top1-mediated DNA relegation (a process that requires the intact 5′-hydroxyl end be aligned with the 3′-end bonded to Top1 for nucleophilic attack of the tyrosyl-phosphoester bond). Free radicals may also cause DSBs by preventing the topoisomerase II (Top2)-mediated DNA religation (Strumberg et al., 2000; Box et al., 2001; Banáth and Olive, 2003; Ohnishi et al., 2009) (Figure 2.2).

UV-INDUCED PYRIMIDINE PHOTOPRODUCTS

UV-B radiation is one of the most important energetic solar components that may lead to the formation of three major classes of DNA lesions, such as cyclobutane pyrimidine dimers (CPDs), pyrimidine 6-4 pyrimidone photoproducts (6-4PPs), and their Dewar isomers. The 6- 4PPs are eagerly converted into their Dewar valence isomers (the third type of environmentally relevant DNA photoproducts induced by solar radiation) upon exposure to UV-B or UV-A radiation that may further undergo reversion to the 6-4PPs upon exposure to short wavelength containing UV radiation. Two adjacent cytosines are considered as mutation hotspots of UVB and UVC radiations.

It has been found that T-T and T-C sequences are more photoreactive than C-T and C-C sequences. The diastereoisomers of pyrimidine dimers (Figure 2.3) can be observed in free solution that differ in the orientation of the two pyrimidine rings relative to the cyclobutane ring, and on the relative orientations of the C5–C6 bonds in each pyrimidine base. The incidence of *trans-syn* isomer in single-stranded or denatured DNA is more common because of the increased flexibility of the DNA backbone. A few CPD lesions (i.e., *cissyn* or *trans-anti* isomers) can also be detected in aqueous conditions (Rastogi et al., 2010; Tewari et al., 2012).

UV-INDUCED PURINE PHOTOPRODUCTS

The extent of adenine-containing photoproduct (A-T) is very low (1×10^{-5} in native DNA) but these lesions may contribute to the biological effects of UV radiation especially at 254 nm in view of the fact that the A-T adduct has been shown to be mutagenic. Photodimerization of adenine (A) involves the cycloaddition of N7-C8 double bond of the 5_-A across the C6 and C5 positions of the 3_-A and generates a very unstable azetidine intermediate. This intermediate photoproduct undergoes competing reaction pathways to form two distinct adenine photoproducts such as adenine dimer (A=A) and P⸚orschke photoproduct (pdApdA or poly(dA)). Conversion of both of these photoproducts into 4,6-diamino-5-guanidinopyrimidine (DGPY) and 8- (5-aminoimidazol-4-yl) adenine (8-AIA), respectively, can be detected from individual acid hydrolysates of UV-irradiated polynucleotides and DNA. It has been found that complexing of UV-irradiated poly(dA)-poly(dT) effectively reduces the formation of A=A photoproduct (Bose et al., 1983; Bose and Davies, 1984).

UV-induced ROS acts as a powerful oxidant that may cause oxidative DNA damage. A number of oxidation products of purine bases such as 8-oxo-7,8-dihydroguanyl (8-oxoGua), 8-oxo-Ade, 2,6-diamino- 4-hydroxy-5-formamidoguanine (FapyGua), FapyAde, and oxazolone have been reported to be formed upon exposure of DNA to UV radiation. Overall, it has been concluded that UV-induced DNA lesions such as CPDs, 6-4PPs, abasic site, strand breaks, and oxidative products are the predominant and most persistent lesions, which if not repaired may cause severe structural distortions in the DNA molecule, thereby affecting the important cellular processes such as DNA replication and transcription, compromising cellular viability and functional integrity and ultimately leading to mutagenesis, tumorigenesis, and cell death (Britt, 2004; Cadet and Wagner, 2013).

HEAT STRESS-INDUCED SINGLE-STRANDED DNA BREAKS (SSBS)

Heat stress not only inhibits DNA repair systems, but can also act as a DNA-damaging agent. It is known that heat stress can lead to the accumulation of 8-oxoguanine, deaminated cytosine, and apurinic DNA sites (AP-sites) in a cell (Bruskov, 2002). Heat stress-induced inhibition of excision repair systems passively accumulated DNA damage, as well as single-stranded DNA breaks (SSBs) in the cell. It has been shown that hyperthermia can inhibit DNA replication: heat stress leads to either a slowing-down or arrest of replication forks, depending on the temperature and cell line (Velichko et al., 2012). Recently, the mechanism of heat stress-induced SSBs has been identified, that heat stress induces SSBs by inhibition of DNA topoisomerase I (top1), an enzyme that relaxes DNA supercoils by introducing temporary SSB into DNA (Velichko et al., 2015). The catalytic cycle of top1 includes cleavage of one DNA strand, accompanied by formation of an intermediate complex consisting of the enzyme covalently bound to the DNA.

Heat stress-induced formation of SSBs is mainly observed in the S-phase of the cell cycle, because the main function of top1 is to resolve topological problems that occur during DNA replication. It could be stated that the sensitivity of non-proliferating cells (terminally differentiated, arrested in G0 phase, etc.) should be significantly reduced in terms of the formation of SSBs. In this regard, it is worth noting that heat stress induction of SSBs is likely to occur not only in the S-phase of the cell cycle. SSBs also form in the G1 and G2 phases, but with very low frequency. It has been observed that the number of SSBs formed due to heat stress in various cell lines directly correlate with the level of top1 expression. It can be concluded that heat stress inhibits the *in vivo* activity of top1 and leads to the formation of covalently bound complexes between the enzyme and the DNA and, as a consequence, formation of SSBs (Kantidze et al., 2016).

HEAT STRESS-INDUCED DOUBLE-STRANDED DNA BREAKS (DSBS)

Heat stress-induced SSB is a source of DSB formation. These DSBs have several interesting features: they are specific to the S-phase of the cell cycle and occur in the cell not immediately after the heat stress, but rather 3–6 hours later. These delayed DSBs occur due to the collision of replication forks which were re-started after heat stress-induced arrest, with SSBs, resulting from top1 inhibition. Slow kinetics of the formation of these DSBs is associated with heat stress-induced inhibition of DNA replication, on the one hand, and inhibition of the transcription process, on the other hand (Velichko et al., 2015). Active transcription process is required for the detection and subsequent removal of the top1 complex covalently bound to DNA, resulting in SSB unmasking and the possibility of their collision with replication forks (Lin et al., 2008; Lin et al., 2009).

The mechanism of delayed DSB formation under heat stress conditions implies that heat stress can induce phosphorylation of H2AX histone, which is one of the first events in the processes of DSB recognition and repair. However, interpretation of these results is quite contradictory: some researchers have stated that γH2AX foci mask heat-induced DSBs; others believe that heat shock itself does not lead to DNA damage and, in this case, γH2AX is a by-product of the cellular response to stress. Recently, it has been proved that hyperthermia can provoke the formation of DSBs. This was confirmed using two independent approaches: comet assay and labeling of DNA ends with terminal deoxynucleotidyl transferase. However, heat stress induces DSBs only in G1- and G2-phase cells. These DSBs are marked by ATM dependent phosphorylation of H2AX. Interestingly, other repair factors, such as the 53BP1 protein, are not attracted to γH2AX foci immediately after exposure to hyperthermia (Petrova et al., 2016). The fact that heat stress can reduce the genotoxic potential of top2 poisons is also indicative of the influence of hyperthermia on this enzyme (Nitiss, 2009). There are two isoforms of top2, and expression of one of them depends on the stage of the cell cycle (Kimura et al., 1994; Goswami et al., 1996). These dynamics of expression could easily explain the dependence of DSB induction on the cell cycle phase.

CONCLUSION

The cellular DNA which is responsible for regulating all its biological activities has been shown to be damaged by several physical (radiations and heat) and chemical (anthropogenic compounds, heavy metals, carcinogens and free radicals) factors. However, as discussed above, the mechanisms of DNA damage by any of these environmental factors are distinct. The radiations (UVR, IR, X-ray) and the heat stress are capable of introducing DNA lesions via formation of cyclobutane pyrimidine dimers, 6-4 photoproducts, SSBs, and DSBs, the latter being highly lethal due to the loss of some important genetic elements.

REFERENCES

Banáth, J.P., Olive, P.L. 2003. Expression of phosphorylated histone H2AX as a surrogate of cell killing by drugs that create DNA double-strand breaks. *Cancer Research* 63(15); 4347–4350.

Batista, L.F.Z., Kaina, B., Meneghini, R., Menck, C.F.M. 2009. How DNA lesions are turned into powerful killing structures: Insights from UV-induced apoptosis. *Mutation Research* 681(2–3); 197–208.

Bose, S.N., Davies, R.J.H. 1984. The photoreactivity of T-A sequences in oligodeoxyribonucleotides and DNA. *Nucleic Acids Research* 12(20); 7903–7914.

Bose, S.N., Davies, R.J.H., Sethi, S.K., McCloskey, J.A. 1983. Formation of an adenine-thymine photoadduct in the deoxydinucleoside monophosphate d(TpA) and in DNA." *Science* 220(4598); 723–725.

Box, H.C., Dawidzik, J.B., Budzinski, E.E. 2001. Free radical-induced double lesions in DNA. *Free Radical Biology and Medicine* 31(7); 856–868.

Britt, A.B. 2004. Repair of DNA damage induced by solar UV. *Photosynthesis Research* 81(2); 105–112.

Bruskov, V.I., Malakhova, L.V., Masalimov, Z.K., Chernikov, A.V. 2002. Heat-induced formation of reactive oxygen species and 8-oxoguanine, a biomarker of damage to DNA. *Nucleic Acids Research* 30(6); 1354–1363.

Cadet, J., Wagner, J.R. 2013. DNA base damage by reactive oxygen species, oxidizing agents, and UV radiation. *Cold Spring Harbor Perspectives in Biology* 5(2). pii:a012559.

Dinant, C., Houtsmuller, A.B., Vermeulen, W. 2008. Chromatin structure and DNA damage repair. *Epigenetics and Chromatin* 1(1); 9. doi: 10.1186/1756-8935-1-9.

Dunkern, T.R., Kaina, B. 2002. Cell proliferation and DNA breaks are involved in ultraviolet light induced apoptosis in nucleotide excision repair-deficient Chinese hamster cells. *Molecular Biology of the Cell* 13(1); 348–361.

Goswami, P.C., Roti, J.L., Hunt, C.R. 1996. The cell cycle-coupled expression of topoisomerase IIalpha during S phase is regulated by mRNA stability and is disrupted by heat shock or ionizing radiation. *Molecular and Cellular Biology* 16(4); 1500–1508.

Halliwell, B., Gutteridge, J. 2007. *Free Radicals in Biology and Medicine*, 4th edition. Oxford University Press, Oxford.

Iliakis, G., Wu, W., Wang, M. 2008. DNA double strand break repair inhibition as a cause of heat radiosensitization: re-evaluation considering backup pathways of NHEJ. *International Journal of Hyperthermia* . 24(1); 17–29.

Kantidze, O.L., Velichko, A.K., Luzhin, A.V., Razin, S.V. 2016. Heat stress-induced DNA damage. *Acta Naturae* 8(2); 75–78.

Kimura, K., Saijo, M., Ui, M., Enomoto, T., Ui, M., Enomoto, T. 1994. Identification of the nature of modification that causes the shift of DNA topoisomerase II beta to apparent higher molecular weight forms in the M phase. *Journal of Biological Chemistry* 269(40); 24523–24526.

Limoli, C.L., Giedzinski, E., Bonner, W.M., Cleaver, J.E. 2002. UV-induced replication arrest in the xeroderma pigmentosum variant leads to DNA doublestrand breaks, γ-H2AX formation, and Mre11 relocalization. *Proceedings of the National Academy of Sciences of the United States of America* 99(1); 233–238.

Lin, C.P., Ban, Y., Lyu, Y.L., Desai, S.D., Liu, L.F. 2008. A ubiquitin-proteasome pathway for the repair of topoisomerase I-DNA covalent complexes. *Journal of Biological Chemistry* 283(30); 21074–21083.

Lin, C.P., Ban, Y., Lyu, Y.L., Liu, L.F. 2009. Proteasome-dependent Processing of Topoisomerase I-DNA Adducts into DNA Double Strand Breaks at Arrested Replication Forks. *Journal of Biological Chemistry* 284(41); 28084–28092.

Minamoto, T., Mai, M., Ronai, Z. 1999. Environmental factors as regulators and effectors of multistep carcinogenesis. *Carcinogenesis* 20(4); 519–527.

Nitiss, J.L. 2009. Targeting DNA topoisomerase II in cancer chemotherapy. *Nature Reviews. Cancer* 9(5); 338–350.

Ohnishi, T., Mori, E., Takahashi, A. 2009. DNA double-strand breaks: Their production, recognition, and repair in eukaryotes. *Mutation Research* 669(1–2); 8–12.

Petrova, N.V., Velichko, A.K., Razin, S.V., Kantidze, O.L. 2016. Early S-phase cell hypersensitivity to heat stress. *Cell Cycle* 15(3); 337–344.

Rastogi, R.P., Singh, S.P., Häder, D.P., Sinha, R.P. 2010. Detection of reactive oxygen species (ROS) by the oxidant-sensing probe 2',7'-dichlorodihydrofluorescein diacetate in the cyanobacterium *Anabaena variabilis* PCC 7937. *Biochemical and Biophysical Research Communications* 397(1); 603–607.

Strumberg, D., Pilon, A.A., Smith, M., Hickey, R., Malkas, L., Pommier, Y. 2000. Conversion of topoisomerase I cleavage complexes on the leading strand of ribosomal DNA into 5'-phosphorylated DNA double-strand breaks by replication runoff. *Molecular and Cellular Biology* 20(11); 3977–3987.

Takahashi, A., Ohnishi, T. 2005. Does γH2AX foci formation depend on the presence of DNA double strand breaks? *Cancer Letters* 229(2); 171–179.

Tewari, A., Sarkany, R.P., Young, A.R. 2012. UVA1 induces cyclobutane pyrimidine dimers but not 6–4 photoproducts in human skin *in vivo. Journal of Investigative Dermatology* 132(2); 394–400.

Valko, M., Leibfritz, D., Moncol, J., Cronin, M.T.D., Mazur, M., Telser, J. 2007. Free radicals and antioxidants in normal physiological functions and human disease. *International Journal of Biochemistry and Cell Biology* 39(1); 44–84.

Velichko, A.K., Petrova, N.V., Kantidze, O.L., Razin, S.V. 2012. Dual effect of heat shock on DNA replication and genome integrity. *Molecular Biology of the Cell* 23(17); 3450–3460.

Velichko, A.K., Petrova, N.V., Razin, S.V., Kantidze, O.L. 2015. Mechanism of heat stress-induced cellular senescence elucidates the exclusive vulnerability of early S-phase cells to mild genotoxic stress. *Nucleic Acids Research* 43(13); 6309–6320.

Effects of chemical environmental factors (both natural and anthropogenic) on DNA and their modes of action. A. Organic factors (pesticides, dyes) B. Inorganic Factors (Heavy metals)

3

INTRODUCTION

Environmental contaminants and toxicants are widespread throughout the world. They can include various materials and chemicals: from by-products of combustion to contaminating trace metals and residual organic compounds used in daily life (Wuana and Okieimen, 2011). Some of the genetic changes occur naturally when DNA is replicated during the process of cell division. But others are the result of environmental exposures that damage DNA. These exposures may include substances such as chemicals in tobacco smoke, asbestos, beryllium, heavy metals (silica lead, arsenic, mercury, cadmium), and organic agents. Potentially toxic chemical compounds may cause genetic mutations or genomic damage, through direct DNA damage or through indirect cellular or physiological effects. Many environmental compounds (possible carcinogens) can lead to cellular stress, directly resulting in alterations of gene expression (Casey et al., 2015; Katre, 2016). The stock of chemical agents has been identified to cause DNA damage through generation of double-strand breaks (DSBs). These chemicals inhibit the biochemical functions of certain key proteins from different DNA repair pathways. On the other hand, the pesticides being indiscriminately used in farming and health practices have been observed to induce mutation, DNA damage, and chromosomal alterations. Copper exists in the nucleus and is closely associated with chromosomes and DNA bases. It has been reported that double-strand breaks of DNA can be activated by copper in association with the pesticides such as fluoxastrobin and imazamox. Industrial chemicals, such as fluoroalkyl substances, brominated flame retardants and butadiene emitted from rubber factories, dichloroethane (an industrial solvent), vinyl chloride, and hydrogen chloride from plastic-manufacturing plants, are reported to seriously damage DNA since they affect the repair machinery involved in fixing DNA aberrations. Food additives, such as coloring agents Red no. 40 (Allura Red AC), Yellow no. 5 (Tetrazine) and Yellow no. 6 (sunset yellow), have been reported to act as carcinogenic agents. Some nitrate and nitrite chemicals such as food preservatives used against bacterial growth are reported as highly toxic and carcinogenic agents. In addition, some cross-linking agents such as mitomycin C and cisplatin, alkylating agents, aromatic compounds, fungal and bacterial toxins are also reported to have DNA-damaging potential (Singh and Sharma, 2019).

There are many mechanisms through which tissue-wide changes could contribute to tumor formation. These include direct modification of gene expression, epigenetic alterations and chromosomal aberrations. DNA double-strand breaks occur largely as a result of damage caused by ionizing radiation. In the study, 15 chemicals with known testicular toxicity were selected from different categories of use. Genotoxic as well as nongenotoxic compounds were chosen. Most of these compounds (acrylamide, acrylonitrile, aflatoxin Bi, Cd2+, Cr6+, cisplatin, DBCP, 1,3-DNB, EDB, styrene oxide, Warn, and thiotepa), have been shown to interact with DNA either spontaneously or after metabolic activation. There is one chemical called benomyl (a systemic benzimidazole fungicide) that has been reported to cause DNA damage by indirect mechanisms. On the other hand, there are some non-genotoxic testicular toxicants such as chlordecone and methoxychlor that are postulated to act via a hormonal mechanism (Bjørge et al., 1996).

The present chapter highlights the impact of chemical toxicants on the genome of organisms. This also includes the general mode of interactions of toxins with DNA.

ORGANIC FACTORS AND THEIR ACTION ON DNA

Exposure to varying amounts and types of organic contaminants is linked to impacts on human health, and the fields of environmental health and toxicology have focused on bettering the understanding of how these contaminants impact biological processes so that their health impacts can be attenuated or prevented. The organic factors may include organic solvents, persistent organic pollutants (polychlorobiphenyls (PCBs), polychlorodibenzo-para-dioxines (PCDD), and polychlorodibenzofuranes (PCDF)), plastic exudates (bisphenol A (BPA) and phthalates), pesticides, brominated flame retardants (BFRs), and polycyclic aromatic hydrocarbons (PAHs). In the early 2000s, investigators began to study how environmental exposures and factors may work at the cellular level not only through genotoxic mechanisms but also through influencing genomic regulation through epigenetic mechanisms (Rhind, 2009; Wuana and Okieimen, 2011; Thompson and Darwish, 2019). Exposure to environmental toxins and pollutants has been responsible for epigenetic modifications. Epigenetics refers to the control of gene expression via mechanisms not directly related to the DNA coding sequence. As a result, all cells in an organism have very different phenotypes despite having the same genome. Epigenetics modulate and regulate gene expression through various epigenomic 'marks', the term given to chemical compounds added to DNA or histone proteins and recognized by enzymes that either lay down or remove the specific mark. These marks change the spatial conformation of chromatin: either compacting it, thereby preventing the binding of transcription factors to the DNA, or opening it, allowing transcription factor binding and usually upregulating cellular processes (Tiffon, 2018; Zitallab et al., 2019). DNA damage may also result from exposure to PAHs. PAHs are potent, ubiquitous atmospheric pollutants commonly associated with oil, coal, cigarette smoke, and automobile exhaust fumes. A common marker for DNA damage due to PAHs is Benzo(a)pyrene diol epoxide (BPDE). BPDE is found to be very reactive and is known to bind covalently to proteins, lipids, and guanine residues of DNA to produce BPDE adducts. If left unrepaired, BPDE-DNA adducts may lead to permanent mutations resulting in cell transformation and ultimately tumor development (Epe, 2002; Jackson and Bartek, 2009).

In a study of gasoline station attendants and traffic police officers, airborne benzene exposure was shown to be associated with a significant reduction in LINE-1 and Alu methylation in peripheral blood DNA (Beceren et al., 2016). These findings show that benzene exposure at relatively low levels may induce altered DNA methylation reproducing the aberrant epigenetic patterns found in malignant cells. Perera et al. (2009) published an exploratory study that used methylation sensitive restriction fingerprinting to analyze umbilical cord white blood cell DNA of 20 children exposed to PAHs. Over 30 DNA sequences were identified whose methylation status was dependent on the level of maternal PAH exposure. In another study, the epidemiologists and environmental health researchers began to investigate the impact of environmental exposure on epigenetic mechanisms in human populations. When considering molecular mechanisms through which long-term programming can occur, epigenetic mechanisms become readily apparent as the key players mediating these effects. Distinct cellular phenotypes are the result of specific gene expression events that are controlled through epigenetic mechanisms. Thus, the entire process of lineage-dependent differentiation characterizing *in utero* development is the result of a complex series of cell-specific epigenetic events, including re-programming of epigenetic patterns such as DNA methylation in a cell-specific context. This differentiation occurs as a series of crucial windows in early fetal development, and the epigenetic mechanisms responsible for differentiation can be influenced by environmental factors. Following fertilization, there is a rapid wave of epigenomic changes. These include changes in the character and modification of histones, particularly those in the paternal germline as the sperm protamines are replaced by histones present in the oocyte, leading to a relative enrichment of histone variant H3.3, and depletion in modifications such as H3K9me2, H3K9me3, and H3K27me3 (Puschendorf et al., 2008).

This replacement is followed immediately by loss of genomic DNA methylation in the paternal genome, occurring through an active demethylation process (Mayer et al., 2000; Oswald et al., 2000). The maternal genome also undergoes demethylation, although through a passive process related to an absence of the maintenance DNA methyltransferase Dnmt1 (Ratnam et al., 2002), and potentially to the presence of the oocyte derived protein PGC7/Stella (Nakamura et al., 2007). Importantly, this demethylation occurs incompletely, as particular genomic regions, including regions involved in genomic imprinting, have been identified to maintain some degree of DNA methylation (Borgel et al., 2010; Smallwood et al., 2011). At least partly, this demethylation is responsible for the totipotency of cells in the morula, while reestablishment of appropriate epigenetic patterning has already begun by the blastocyst stage, allowing for the differentiation of the trophectoderm, which will develop into extraembryonic tissue from the inner cell mass that will form the fetus (Santos et al., 2010). From this point, DNA methylation is reestablished and is responsible for cell lineage-dependent differentiation leading to the formation of each of the fully differentiated cells of the offspring (Borgel et al., 2010).

INORGANIC FACTORS AND THEIR ACTIONS ON DNA

In environmental studies, the flexibility of epigenetic states has generated growing interest in evaluating whether environmental exposures can modify epigenetic states, including DNA methylation and histone modifications. Studies of DNA methylation and histone modification in relation to environmental exposures to potentially toxic chemicals have not been systematically carried out. The metallic elements including arsenic, cadmium, chromium, lead, and mercury are considered systemic toxicants that are known to induce multiple organ damage, even at lower levels of exposure. Their multiple industrial, domestic, agricultural, medical, and technological applications have led to their wide distribution in the environment; raising concerns over their potential effects on human health and the environment. Metal ions have been found to interact with cell components such as DNA and nuclear proteins, causing DNA damage and conformational changes that may lead to cell cycle modulation, carcinogenesis, or apoptosis (Hou et al., 2012). Several studies have demonstrated that ROS production and OS play a key role in the toxicity and carcinogenicity of metals such as arsenic, cadmium, chromium, lead, and mercury. Because of their high degree of toxicity, these five elements rank among the priority metals that are of great public health significance. They are all systemic toxicants that are known to induce multiple organ damage, even at lower levels of exposure (Wang et al., 2016).

Various hypotheses have been proposed to explain the carcinogenicity of inorganic arsenic. Nevertheless, the molecular mechanisms by which this arsenical induces cancer are still poorly understood. The reports from the study have indicated that inorganic arsenic does not act through classic genotoxic and mutagenic mechanisms, but rather may be a tumor promoter that modifies signal transduction pathways involved in cell growth and proliferation (Stýblo et al., 2002). Based on the comet assay, it has been reported that arsenic trioxide induces DNA damage in human lymphophytes and also in mice leukocytes. Although much progress has been recently made in the area of arsenic's possible mode(s) of carcinogenic action, a scientific consensus has not yet reached. A recent review has presented nine different possible modes of action of arsenic carcinogenesis: induced chromosomal abnormalities, oxidative stress, altered DNA repair, altered DNA methylation patterns, altered growth factors, enhanced cell proliferation, promotion/progression, suppression of p53, and gene amplification (Tchounwou et al., 2012).

Presently three modes such as chromosomal abnormality, oxidative stress, and altered growth factors of heavy metal (arsenic) carcinogenesis have shown a degree of positive evidence, both in experimental systems (animal and human cells) and in human tissues. The remaining possible modes of carcinogenic action (progression of carcinogenesis, altered DNA repair, p53 suppression, altered DNA methylation patterns, and gene amplification) do not have as much evidence, particularly from *in vivo* studies with laboratory animals, *in vitro* studies with cultured human cells, or human data from case or population studies. Thus, the mode-of-action studies suggest that arsenic might be acting as a cocarcinogen, a promoter, or a progressor of carcinogenesis (Kitchin, 2001).

The primary mechanism of cadmium toxicity causing cellular damage to the cells primarily concerns the ROS generation resulting in single-strand DNA damage thereby inhibiting the biosynthesis of key biomolecules i.e. the nucleic acids and proteins. The *in vitro* studies have indicated the cytotoxic effects of cadmium at 0.110 mM concentrations. They have shown in the cell culture systems that cadmium induces free radicals and induce DNA damage. They have observed that all the carcinogenic metals tested (arsenic, cadmium, chromium, and nickel) displayed DNA-damage potential via inducing base pair mutations, deletion, or attack by free radical oxygen species on DNA (Tchounwou et al., 2012).

Under the physiological conditions, another heavy metal i.e. Cr (VI) enters several cell types. Cr (VI) can get reduced by many reducing agents such as ascorbic acid, glutathione (GSH) reductase, GSH, and hydrogen peroxide (H_2O_2) to produce reactive intermediates such as Cr(V), Cr (IV), Cr (III), hydroxyl radicals, and thiyl radicals. These chemical species are known to attack biological macromolecules such as DNA, membrane lipids, and proteins. These chemical species disrupt integrity of cells and their functions. The results from the epidemiological investigations have reflected the adverse effects caused by Cr (VI) on the health of humans in the form of respiratory cancers in workers who got occupational exposure to the compounds containing Cr (VI). In addition to it, the breaks in DNA strand(s) in the peripheral blood lymphocytes as well as excretion of the products of lipid peroxidation in the urine have also been observed in the chromium-exposed occupants. These evidences support that exposure of humans to the oxidized chemical form of Cr(VI) is capable of inducing considerable level of adverse effects in humans. The underlying mechanism of the toxic effects of cadmium exposure has been found to be via free radical production followed by the oxidative damage which finally culminates into the genotoxic effects such as chromosomal abnormalities and breaks in the DNA strand(s) (Yao et al., 2008; Das et al., 2015; Xu et al., 2019).

Moreover, some studies have also indicated the biological relevance of the non-oxidative mechanisms involved in Cr(VI) mediated carcinogenesis (Shi and Dalal, 1989; Liu and Shi, 2001). The results of these studies have indicated cytotoxicity of Cr(VI). They also showed that Cr(VI) was able to induce DNA damage and hence it could cause chromosomal abnormalities, fragmentation and breaks of DNA as well as oxidative stress in the experimental animals like Sprague-Dawley rats and human hepatic carcinoma cells (Patlolla et al., 2009a; Patlolla et al., 2009b). The results of the experiments carried out by other workers in the kidney and liver of *Carassius auratus*, commonly known as a gold fish, have also demonstrated that Cr(VI) may exert adverse effects on the biochemical parameters showing genotoxic and histopathologic impacts (Velma and Tchounwou, 2010).

A series of recent studies have demonstrated lead-induced toxicity and apoptosis in human cancer cells. Lead was found to be involved in several cellular and molecular processes including induction of cell death and oxidative stress, transcriptional activation of stress genes, DNA damage, and externalization of phosphatidylserine and activation of caspase-3. The *in vitro* and *in vivo* studies have indicated that lead causes genetic damage through various indirect mechanisms that include inhibition of DNA synthesis and repair, oxidative damage, and interaction with DNA-binding proteins and tumor suppressor proteins. However, *in vitro* studies suggest that the susceptibility to DNA damage exists as a result of cellular exposure to both lead and mercury (Yedjou et al., 2010).

CONCLUSION

There is a huge list of diverse range of materials and anthropogenic chemicals containing compounds with heavy metals, which are continuously being manufactured by many industries and continuously released into the environment in different forms. The interaction of humans and other living beings with these physical and chemical factors is leading to significant decrease in the quality of their lives in terms of emergence of several diseases including asthma, hypertension, and cancer. Many of them possess genotoxic potential and damage their DNA through different pathways as discussed above. Exposure of living systems to varying concentrations of these anthropogenic chemicals so often generate reactive oxygen species which not only cause breaks in DNA strand(s) but also induce necrosis and lesions in tissues of different organs leading to rapid aging, reduction in immunity, and early death.

REFERENCES

Beceren, A., Akdemir, N., Omurtag, Z.G., Tatlıpınar, M.E., Şardaş, S. 2016. DNA damage in gasoline station workers caused by occupational exposure to petrol vapour in turkey. *ACTA Pharmaceutica Sciencia* 54(1); 1–10.

Bjørge, C., Brunborg, G., Wiger, R., Holme, J.A., Scholz, T., Dybing, E., Spiderlund, E.J. 1996. A comparative study of chemically induced dna damage in isolated human and rat testicular cells. *Reproductive Toxicology* 10(6); 509–519.

Borgel, J., Guibert, S., Li, Y., Chiba, H., Schübeler, D., Sasaki, H., Forné, T., Weber, M., 2010. Targets and dynamics of promoter DNA methylation during early mouse development. *Nature Genetics* 42; 1093–1100.

Casey, S.C., Vaccari, M., Al-Mulla, F., Al-Temaimi, R., Amedei, A., Barcellos-Hoff, M.H., Brown, D.G., Chapellier, M., Christopher, J., Curran, C.S., Forte, S., Hamid, R.A., Heneberg, P., Koch, D.C., Krishnakumar, P.K., Laconi, E., Maguer-Satta, V., Marongiu, F., Memeo, L., Mondello, C., Raju, J., Roman, J., Roy, R., Ryan, E.P., Ryeom, S., Salem, H.K., Scovassi, A.I., Singh, N., Soucek, L., Vermeulen, L., Whitfield, J.R., Woodrick, J., Colacci, A., Bisson, W.H., Felsher, D.W. 2015. The effect of environmental chemicals on the tumor micro-environment. *Carcinogenesis* 36(Supplement 1); S160–183.

Das, J., Sarkar, A., Sil, P.C. 2015. Hexavalent chromium induces apoptosis in human liver (HepG2) cells via redox imbalance. *Toxicology Reports* 2; 600–608.

Epe, B. 2002. Role of endogenous oxidative DNA damage in carcinogenesis: What can we learn from repair-deficient mice? *Biological Chemistry* 383(3–4); 467–475.

Hou, L., Zhang, X., Wang, D., Baccarelli, A. 2012. Environmental chemical exposures and human epigenetics. *International Journal of Epidemiology* 41(1); 79–105.

Jackson, S.P., Bartek, J. 2009. The DNA-damage response in human biology and disease. *Nature* 461(7267); 1071–1078.

Katre, S.D. 2016. Types, sources and effects of chemical toxicants: An overview. *Der Pharmacia Sinica* 7; 40–45.

Kitchin, K.T. 2001. Recent advances in arsenic carcinogenesis: Modes of action, animal model systems, and methylated arsenic metabolites. *Toxicology Applied Pharmacology* 172(3); 249–261.

Liu, K.J., Shi, X. 2001. In vivo reduction of chromium (VI) and its related free radical generation July. *Molecular and Cellular Biochemistry* 222(1–2); 41–47.

Mayer, W., Niveleau, A., Walter, J., Fundele, R., Haaf, T. 2000, Demethylation of the zygotic paternal genome. *Nature* 403; 501–502.

Nakamura, T., Arai, Y., Umehara, H. et al. 2007. PGC7/Stella protects against DNA demethylation in early embryogenesis. *Nature Cell Biology* 9; 64–71.

Oswald, J., Engemann, S., Lane, N., Mayer, W., Olek, A., Fundele, R., Dean, W., Reik,W., Walter, J. 2000, Active demethylation of the paternal genome in the mouse zygote. *Current Biology* 10(8);475–478.

Patlolla, A., Barnes, C., Field, J., Hackett, D., Tchounwou, P.B. 2009a. Potassium dichromate-induced cytotoxicity, genotoxicity and oxidative stress in human liver carcinoma (HepG2) cells. *International Journal of Environmental Research and Public Health* 6(2); 643–653.

Patlolla, A., Barnes, C., Yedjou, C., Velma, V., Tchounwou, P.B. 2009b. Oxidative stress, DNA damage and antioxidant enzyme activity induced by hexavalent chromium in Sprague Dawley rats. *Environmental Toxicology* 24(1); 66–73.

Perera, F., Tang, W.Y., Herbstman, J., Tang, D., Levin, L., Miller, R., Ho, S.M. 2009. Relation of DNA methylation of 5'-CpG island of ACSL3 to transplacental exposure to airborne polycyclic aromatic hydrocarbons and childhood asthma. *PLOS ONE* 4(2); e4488.

Puschendorf, M., Terranova, R., Boutsma, E., Mao, X., Isono, K., Brykczynska, U., Kolb, C., Otte, A.P., Koseki, H., Orkin, S.H., et al. 2008. PRC1 and Suv39h specify parental asymmetry at constitutive heterochromatin in early mouse embryos. *Nature Genetics* 40; 411–420.

Ratnam, S., Mertineit, C., Ding, F., Howell, C.Y., Clarke, H.J., Bestor, T.H., Chaillet, J.R., Trasler, J.M. 2002. Dynamics of Dnmt1 methyltransferase expression and intracellular localization during oogenesis and preimplantation development. *Developmental Biology* 245; 304–314.

Rhind, S.M. 2009. Anthropogenic pollutants: A threat to ecosystem sustainability? *Philosophical Transactions of the Royal Society of London B Biological Sciences* 364(1534); 3391–3401.

Santos, J., Filipe Pereira, C., Di-Gregorio, A. et al. 2010. Differences in the epigenetic and reprogramming properties of pluripotent and extra-embryonic stem cells implicate chromatin remodelling as an important early event in the developing mouse embryo. *Epigenetics & Chromatin* 3; 1

Shi, X., Dalal, N.S. 1989. Chromium (V) and hydroxyl radical formation during the glutathione reductase-catalyzed reduction of chromium (VI). *Biochemical and Biophysical Research Communications* 163(1); 627–634.

Singh, N., Sharma, B. 2019. Role of toxicants in oxidative stress mediated DNA damage and protection by phytochemicals. *EC Pharmacology and Toxicology* 7(5); 325–330.

Smallwood, S.A., Tomizawa, S., Krueger, F., Ruf, N., Carli, N., Segonds-Pichon, A., Sato, S., Hata, K., Andrews, S.R., Kelsey, G. 2011. Dynamic CpG island methylation landscape in oocytes and preimplantation embryos. *Nature Genetics* 43:811–814.

Stýblo, M., Drobná, Z., Jaspers, Ilona, Lin, S., Thomas, D.J. 2002. The role of biomethylation in toxicity and carcinogenicity of arsenic: A research update Environmental Health Perspectives. *Molecular Mechanisms of Metal Toxicity and Carcinogenicity* 110(Supplement 5); 767–771.

Tchounwou, P.B., Yedjou, C.G., Patlolla, A.K., Sutton, D.J. 2012. Heavy metal toxicity and the environment. *Experientia Supplementum* 101; 133–164.

Thompson, L.A., Darwish, W.S. 2019. Environmental chemical contaminants in food: Review of a global problem. *Journal of Toxicology* 2019: 2345283.

Tiffon, C. 2018. The impact of nutrition and environmental epigenetics on human health and disease. *International Journal of Molecular Sciences* 19(11). pii: E3425.

Velma, V., Tchounwou, P.B. 2010. Chromium-induced biochemical, genotoxic and histopathologic effects in liver and kidney of goldfish, Carassius auratus. *Mutation Research* 698(1–2); 43–51.

Wang, L., Wise, J.T., Zhang, Z., Shi, X. 2016. Progress and prospects of reactive oxygen species in metal carcinogenesis. *Current Pharmacology Reports* 2(4); 178–186.

Wuana, R.A., Okieimen, F.E. 2011. Heavy metals in contaminated soils: A review of sources, chemistry, risks and best available strategies for remediation. *International Scholarly Research Network* 2011; 1–20.

Xu, J., Zhao, M., Pei, Lu, Zhang, R., Liu, X., Wei, L., Yang, M., Xu, Q. 2019. Oxidative stress and DNA damage in a long-term hexavalent chromium-exposed population in North China: A cross-sectional study. *BMJ Open* 8(6); e021470.

Yao, H., Guo, L., Jiang, B., Luo, J., Shi, X. 2008. Oxidative stress and chromium (VI) carcinogenesis. *Journal of Environmental Pathology Toxicology and Oncology* 27(2); 77–88.

Yedjou, C.G., Milner, J.N., Howard, C.B., Tchounwou, P.B. 2010. Basic apoptotic mechanisms of lead toxicity in human leukemia (HI-60) cells. *International Journal of Environmental Research and Public Health* 7(5); 2008–2017.

Zitallab, P.N., Baruah, K., Vanrompay, D., Bossier, P. 2019. Can epigenetics translate environmental cues into phenotypes? *Science of The Total Environment* 647; 1281–1293.

Biological factors causing DNA damage. A. Plant-based molecules. B. Animal-based molecules

4

INTRODUCTION

Generally, DNA of all living organisms is stable. However, continuous exposure of any living organism to numerous biological environmental factors may cause damage to the genomic DNA, resulting in diverse heritable syndromes, which are transferred from one generation to the next. Biological factors comprise toxins derived from animals and plant sources including their metabolites such as phytochemicals, hormones, and other by-products (Singh and Sharma, 2019). A biotoxin is a poisonous substance produced within living cells or organisms. We have not included here the varied synthetic toxicants industrially manufactured or synthesized using artificial processes. The term 'toxin' was first used by an organic chemist, Ludwig Brieger (1849–1919). Toxins can be small molecules, peptides, or proteins that are capable of causing disease on contact with or absorption by the body tissues and interacting with biological macromolecules such as enzymes or cellular receptors. Toxins vary greatly in their toxicity, ranging from usually minor (such as a bee sting) to almost immediately deadly (such as botulinum toxin) (Lahiani et al., 2017; Zhang et al., 2017; Singh and Sharma, 2018a).

Generally, toxins are metabolic by-products produced into the animals and plants species. Natural compounds such as plant products and human hormones can also act as toxins. These toxins enter into the body of an organism through several routes of exposure such as dermal contact, inhalation, ingestion, injection, or accident. The adverse effects of these chemicals include membrane damage, protein dysfunction, DNA impairment, and disorder of metabolism. They may negatively modulate signaling pathways and cause mutagenicity as well as cell death (Singh and Sharma, 2018).

In addition, some bacterial and viral infections such as *Helicobacter pylori*, human immunodeficiency virus (HIV), and human papillomavirus have been shown to possess DNA-damage potential. A biological cell always stays under a threat of DNA damage by both endogenous and exogenous factors. These toxins enter into the body of an organism through several routes such as dermal contact, inhalation, ingestion, injection, and accident. The plant and animal by-products are reported to cause membrane damage, protein dysfunction, DNA impairment, metabolic disorders, mutagenicity, cancer and cell death, etc. They induce their effects on DNA by distorting its structure via breaking of hydrogen bonds involved in DNA strands stabilization. The types of DNA damage include oxidative damage, hydrolytic damage, and DNA-strand breaks (Wroblewski et al., 2010; Kalisperati et al., 2017).

Exposure of DNA from living systems to different biotoxins is believed to lead to disruption of native structure and function of DNA thereby introducing a variety of genetic disorders. However, only meager information is available in this context. The impact of biotoxins on DNA has been described by only a small number of workers. For instance, Ochratoxins which are a group of mycotoxins produced by some Aspergillus species and some Penicillium species act as potential nephrotoxins and renal carcinogens. Ochratoxin A (OTA) has been reported to cause DNA strand breaks in liver, kidney, and spleen of treated animals and a similar degree of DNA damage was observed in rats treated with Ochratoxin A (OTB). In addition, there are many plants, which might appear innocent and harmless at first glance, but there are many which produce toxins dangerous to humans and other animals. Phytotoxin is a toxin which is only obtained from plant sources. The adverse effects of phytotoxin have been reported to inhibit the physiological and biochemical processes in plants as well as animals (Möbius and Hertweck, 2009). These plant products cause harm to the organisms by indirect and direct modes of exposure. Indirectly, such plants release chemicals into the environment through root exudation, leaching, and volatilization, and passively through decomposition of plant residues, which may further be taken up by organisms through inhalation, dermal contact, and also through ingestion and interrupt organism's growth by interfering with their physiological and biochemical processes (Bender et al., 1998: Bignell et al., 2018). However, the direct exposure concerns the plant components or plant parts directly taken up by organism either with or without other food materials (Wang et al., 2016). This chapter deals with the different aspects of DNA damage induced by varied natural products released by the plants and animals in the environment.

DETECTION OF DNA DAMAGE

In order to detect DNA damage, it is a prerequisite to take up a study to determine the site of injury and mode of DNA damage. Principally, any of the detection methods associated with the molecular structure of DNA can be tried for detecting DNA damage. The detection of DNA damage in any organ or tissue of a living organism is a complex process because of the existence of a very complicated biochemical environment around the genomic DNA. In this context, the study of certain parameters such as aberrance in DNA, cell survival, and cell division are carried out in order to detect and assess the level of DNA. About a decade ago, DNA damage detection technology was developed to ascertain the extent of DNA damage and to study it at the molecular level. The technology confirms the site and the style of damage of DNA due to exposure of an organism to certain physical and/or chemical agents. Popular methods of studying DNA damage include biological methods, spectroscopy, electrophoresis, comet assay analysis, DNA chip, and molecular probe technologies (June, 2010; Shao et al., 2010).

Comet assay, also called a single cell electrophoresis technique (SCGE), has been considered to be a rapid, simple, and sensitive method for detection of DNA damage. This technique has widely been used to detect and identify the break(s) of DNA strand or damage due to DNA cross-linking and single-cell DNA damage (Merk, 1999). For detecting DNA damage, changes in spectrum property can also be considered. Another DNA damage detection technique could be fluorescent spectrophotometry (FSP), which depends on the intrinsic fluorescence properties of some compounds such as polycyclic aromatic hydrocarbons (PAH). In order to detect DNA adducts, other viable methods include fluorescence emission spectrophotometry (FES), synchronous fluorimetric spectrophotometry (SFS), and methods involving cryogenic laser. Fluorescent spectrophotometry (FP) is an extremely highly sensitive method, with the detection limit to be 106~108. Though the synchronous fluorimetric technique is one of the most commonly used methods, it is not suitable for determination of DNA damage caused by non-fluorescent intercalated agents (Vahakangas, 1985).

In recent years, the determination of damage of DNA has been carried out using DNA chip and molecular probes. It is a very rapid and efficient method of DNA damage detection while analyzing mutations or variations in the genome sequence, spectrum analysis of expression of genes, diagnosis of disease, screening of drugs, detection of pathogens, genome sequence related to a disease, changes in copy number, and nucleotide sequence polymorphism detection (Shaon, 1996). The use of the synthetic oligonucleotides as molecular probes can detect the mutation via an analysis of the hybridization signal. This technique could be employed for developing methods towards detection of chemical modification or damage of DNA (Gong, 2007).

Another technique to determine DNA damage could be an immunoassay. The immunology principle concerning the reaction between an antigen and the antibody is utilized to detect the antibody generated against DNA adducts in the injured tissue; the most frequently used methods are the competitive immunoassay, a non-competitive enzyme-linked immunosorbent assay, solid-phase competition methods, or ultra-sensitive enzymatic radioimmunoassay, etc. (Ding et al., 2017). The immunoassay sensitivity for detection of DNA adducts is very high. It could be performed very easily using cultured cells *in vitro*, viz. lymphocytes. The technique is of great benefit as it is quite simple and cost effective and does not need enzomolysis of DNA strand.

MECHANISM OF PLANT-BASED MOLECULES AS DNA-DAMAGING AGENT

Phytotoxins are plant-based compounds acting as toxins. The effects of phytotoxins have been mostly reported in terms of inhibition of the physiological and biochemical processes in the plants as well as animals (Möbius and Hertweck, 2009). Phytotoxins cause harm to organisms through indirect and direct exposures of these agents. Indirectly, plants release toxic chemicals into the environment through root exudation, leaching, and volatilization, and passively through decomposition of plant residues, which may be taken up by other organisms through inhalation, dermal contact, ingestion, and accidental events. Phytotoxins interrupt the growth of organisms by altering their physiological and biochemical processes (Bender et al., 1998; Bignell et al., 2018). However, direct exposure may be caused when the plant components or plant parts are directly taken up through inhalation or contact or ingestion along with other food stuffs (Wang et al., 2016).

There are some plant products which are shown to possess DNA-damaging potential. Recent findings suggest an active role of nicotine, a major alkaloid present in tobacco, as carcinogen. Nicotine present in tobacco or tobacco products exhibits tumor-promoting potential by causing DNA damage in different human epithelial and non-epithelial cells (Yamaguchi, 2019). Sanguinarine, an alkaloid isolated from the wild plant *Argemone mexicana*, has been shown to cause chromosomal aberration, micronucleus formation, and DNA damage by comet assay in mouse model *in vivo* system. Sanguinarine is reported to inhibit the activity of epidermal histidase, leading to increase in the levels of keratin formation and tumor promotion.

An extensive survey of available literature indicates that phytotoxin can induce oxidative stress (OS), which has been considered as a possible mechanism of toxicity and hence it has been a focus of toxicological research these days. Phytotoxins have been shown to induce production of reactive oxygen species (ROS) by altering the balance between oxidants/prooxidants and antioxidants through promoting lipid peroxidation (LPO) and depleting the antioxidative cellular reserves (both the enzymatic and non-enzymatic components) leading to a condition of OS (Duke and Dayan, 2011; Bignell et al., 2013). The range of their impact spans from tissue injury and aging through apoptosis to onset of various known/unknown diseases. However, the exact mechanism(s) of their actions in plant and animal systems has still not been completely known (Singh and Sharma, 2018; Singh and Sharma, 2019; Martin and Frisan, 2020).

MECHANISM OF ANIMAL-BASED MOLECULES AS DNA-DAMAGING AGENTS

Biotoxins are recognized as substances produced by living organisms; they induce toxicity after introduction in other living systems. The living organism can be a plant or an animal. Generally, toxins are the metabolic byproducts of animals and plants. Natural compounds such as plant products and human hormones can also act as toxins. After getting entry into the body of an organism through several routes of exposure such as dermal contact, inhalation, ingestion, injection, or accident, these molecules adversely influence the physiological health of living systems. These chemicals have been observed to cause membrane damage, protein dysfunction, DNA impairment, and disorder of metabolism. They may also negatively modulate signaling pathways and cause mutagenicity as well as cell death (Duke and Dayan, 2011, Thakur et al., 2018).

The mechanisms of oxidative DNA damage in various organisms have not been elucidated properly. However, oxidative DNA damage mediated by Fenton reactions has been reported to be the most acceptable hypothesis (Singh et al., 2020). Free radicals, commonly known as reactive oxygen species, contain one or more unpaired electrons in their outer most orbital. Excessive production of free radicals results in depletion of antioxidants *in vivo* and causes an imbalance between free radicals and the antioxidant defenses of the body, which results in generation of oxidative stress mediated DNA damage. 8-hydroxydeoxyguanosine (8-OHdG) is the most common biomarker of oxidative DNA damage by chemical carcinogens in which oxidation of a specific base, i.e. guanosine in DNA causes increase in the level of hydroxydeoxyguanosine (8-OHdG). These oxidative chemical species may cause deamination of cytosine converting it into uracil or may remove an individual base generating apurinic/apyrimidinic (AP) sites in DNA (Singh and Sharma, 2019, Singh et al., 2020).

Toxins induce their effects by distorting the DNA structure through breakage of hydrogen bonds between two complementary base pairs involved in stabilization of DNA strands. In order to maintain the genome integrity, it is necessary to repair the DNA damage with the help of DNA repair machineries. Any abnormality in DNA repair mechanism can result in genomic instability. The cross-linking agents such as mitomycin C and aromatic compounds, fungal and bacterial toxins, and metabolic products such as free radicals or reactive oxygen/nitrogen species (ROS/RNS) play crucial role in DNA damage. Free radicals induce DNA damage in living organisms by a variety of mechanisms. These include DNA base and sugar products, single- and double-strand breaks, 8,5'-cyclopurine-2'-deoxynucleosides, tandem lesions, clustered sites, and DNA-protein cross-links (Brosh, 2013; Singh and Sharma, 2018).

CONCLUSION

The biotoxins, i.e. toxic molecules produced by living organisms (plants and animals), have been shown to cause DNA damage after entering into the living cells. Different groups of biotoxins have been observed to follow different pathways to disrupt the structure of the genomic DNA. The detection of DNA damage in any organ or tissue of a living organism is a complex process because of existence of a very complicated biochemical environment around the genomic DNA. The popular methods of studying DNA damage include biological methods, comet assay analysis, or single cell electrophoresis technique (SCGE), spectroscopy (fluorescent spectrophotometry [FSP], fluorescence emission spectrophotometry [FES], synchronous fluorimetric spectrophotometry [SFS], and methods involving cryogenic laser), DNA chip, molecular probe technologies, electrophoresis, and immunoassay techniques (for detection of DNA adducts). Nicotine and sanguinarine, the phytotoxins, cause DNA damage, chromosomal aberration, or micronucleus formation in both the epithelial and non-epithelial cells. Sanguinarine is known to inhibit the epidermal histidase activity leading to increase in the chances of tumor promotion. Biotoxins-induced free radical production by generation of oxidative stress was found to be the underlying mechanism of actions of these molecules. Similar mechanism of action of biotoxins of animal origin such as hormones has been reported to cause damage to cellular membrane, protein, and DNA and alter the level of metabolism. These animal-based biotoxins may also adversely modulate signaling pathways, mutagenicity, and cell death.

REFERENCES

Bender, C.L., Palmer, D.A., Penaloza-Vazquez, A., Rangaswamy, V., Ullrich, M. 1998. Biosynthesis and regulation of coronatine, a non-host-specific phytotoxin produced by *Pseudomonas syringae*. In: *Plant-Microbe Interactions*, Biswas, B.B. and Das, H.K. (eds.), Springer, Boston, 321–341.

Bignell, D.R., Cheng, Z., Bown, L. 2018. The coronafacoyl phytotoxins: Structure, biosynthesis, regulation and biological activities. *Antonie Leeuwenhoek* 111(5); 649–666.

Bignell, D.R.D., Fyans, J.K., Cheng, Z. 2013, Phytotoxins produced by plant pathogenic *Streptomyces* species, *Journal of Applied Microbiology* 116; 223–235.

Brosh, R.M. 2013. DNA helicases involved in DNA repair and their roles in cancer. *Nature Review Cancer* 13(8); 542–558.

Ding, Y., Hua, X., Sun, N., Yang, J., Deng, J., Shi, H., et al. 2017. Development of a phage chemiluminescent enzyme immunoassay with high sensitivity for the determination of imidaclothiz in agricultural and environmental samples. *Science of the Total Environment* 609; 854–860.

Duke, S.O., Dayan, F.E. 2011. Modes of action of microbially-produced phytotoxins. *Toxins* 3(8); 1038–1064.

Gong, J., Sturla, S.J. 2007. A synthetic nucleoside probe that discerns a DNA adduct from unmodified DNA. *Journal of American Chemistry Society* 129(16); 4882–4883.

Kalisperati, P., Spanou, E., Pateras, I.S., Korkolopoulou, P., Varvarigou, A., Karavokyros, I., Gorgoulis, V.G., Vlachoyiannopoulos, P.G., Sougioultzis, S. 2017. Inflammation, DNA damage, helicobacter pylori and gastric tumorigenesis. *Frontiers Genetics* 8; 20.

Lahiani, A., Yavin, E., Lazarovici, P. 2017. The molecular basis of toxins' interactions with intracellular signaling via discrete portals. *Toxins (basel)*. 9; E107.

Martin, O.C.C.B., Frisan, T. 2020. Bacterial genotoxin-induced DNA damage and modulation of the host immune microenvironment. *Toxins* 12(2); 1–22.

Merk, O., Speit, G. 1999. Detection of crosslinks with the comet assay in relationship to genotoxicity and cytotoxicity. *Environmental and Molecular Mutagenesis* 33(2); 167–172.

Möbius, N., Hertweck, C. 2009. Fungal phytotoxins as mediators of virulence. *Current Opinion in Plant Biology* 12(4); 390–398.

Shao, J. 2010. DNA chemical damage and its detected. *International Journal of Chemistry* 2; 261–265.

Shaon, D., Smith, J.S., Brown, P.O. 1996. A DNA microarry system for analyzing complex DNA samples using 2-color fluorescent-probe hybridization. *Genome Research* 6(7); 639–643.

Singh, N., Sharma, B. 2018a. Biotoxins mediated DNA damage and role of phyto-chemicals in DNA potection. *Biochemistry Molecular Biological Journal* 4; 1–5.

Singh, N., Sharma, B. 2018b. A brief review on the recent advances in phytotoxin mediated oxidative stress. *Pharmacologia* 9; 85–93.

Singh, N., Sharma, B. 2019. Role of toxicants in oxidative stress mediated DNA damage and protection by phytochemicals. *EC Pharmacology and Toxicology* 7(5); 325–330.

Singh, N., Gupta, V.K., Doharey, P.K., Srivastava, N., Kumar, A., Sharma, B. 2020. A study on redox potential of phytochemicals and their impact on DNA. *Journal of DNA RNA Research* 1(2); 10–22.

Thakur, A., Sharma, V., Thakur, A. 2018. Phytotoxins - A mini review. *Journal of Pharmacognosy and Phytochemistry* 7; 2705–2708.

Vahakangas, K., Haugen, A., Harris, C.C. 1985. An applied synchronous fluorescence spectro- photometric assay to study benzo[a] pyrene-diolepoxide-DNA adducts. *Carcinogenesis* 8; 1109–1115.

Wang, S.F., Guo, C.R., Xu, X.Y., Zhu, L. 2016. Review on biological treatment of cyano-bacterial toxin in natural waters. *Journal of Applied Ecology* 27(5); 1683–1692.

Wroblewski, L.E., Peek, R.M., Wilson, K.T. 2010. Helicobacter pylori and gastric can-cer: Factors that modulate disease risk. *Clinical Microbiology Reviews* 23(4); 713–739.

Yamaguchi, N.H. 2019. Smoking, immunity, and DNA damage. *Translational Lung Cancer Research* 8(Suppl 1); S3–S6.

Zhang, T.Y., Zhao, Y., Li, L., Shen, W. 2017. Toxicological characteristics of ochra-toxin A and its impact on male reproduction. *National Journal of Andrology* 23(8); 757–762.

Synergistic effects of environmental factors on DNA damage.
A. Introduction to Synergism
B. Synergistic effects
C. Mechanism of actions

5

INTRODUCTION TO SYNERGISM

All living organisms in nature are frequently exposed to a mixture of xenobiotics (heavy metals, pesticides, and toxic gases, etc.) simultaneously. Xenobiotic substances have been reported to cause toxicity in key organs of animals and human beings (Omiecinski et al., 2011; Oesch et al., 2014). Therefore, combined interactions between xenobiotic substances as well as xenobiotic and animal systems are very important (Oesch et al., 2014). The synergy is the concept of additive effects. This has also been referred to as non-interaction, and inertism (Greco et al., 1995). An additive effect is generally considered as the baseline effect for synergy detection methods. Theoretically, this effect is expected from the combination of multiple drugs when synergy is not present. Although seemingly simple from its name, the idea of simply adding two, or more, effects together does not accurately reflect what happens realistically. A quick, simple example to show this involves two separate drugs A and B that both exhibit 60% cell death at a saturating dose, for example. Would it be possible to simply add these effects together to get a combined effect of 120% cell death? No, that is not realistically possible (Chou, 2010). The problem of mathematically defining additivity has been the center of controversy among leading researchers of this topic for the last century. However, there are two models that have prevailed and will further be described in the Reference Models section. Along with these prevailing models, there are two additional terms, Loewe Additivity and Bliss Independence (Greco et al., 1995), that are essentially synonymous with the basic definition of additive but exist for these specific reference models. Loewe Additivity has also been referenced as dose additivity and Concentration Addition (Cedergreen, 2014).

Any (significant) deviation from additivity would be classified as synergy or antagonism. It is often agreed that synergy can be defined as a combination effect that is greater than the additive effect expected from good knowledge of the individual drugs. Synergy has also been called super-additivity (Tallarida, 2001), potentiation, augmentation, and supra-additivity (Geary, 2013). The term 'coalism' is also sometimes used to refer to synergy when neither drug nor the drugs in mixtures of more than two chemicals, is effective on its own (Greco et al., 1995). There are also two distinctively termed ideas to describe synergy under the previously mentioned specific models. Those terms are Loewe Synergy and Bliss Synergy; the models from which the terms have been derived would be discussed further in Reference Models.

Antagonism is the opposite of synergy; it occurs when the combined effect of compounds is less than what would be expected. In the biomedical world, it is often considered more of a negative scenario, as many researchers are looking to identify synergistic interactions among compounds for some sort of added therapeutic effect. However, in a toxicological sense, it may be beneficial to have an antagonistic effect in a mixture of chemicals. Antagonism has also been named subadditivity (Tallarida, 2001), infra-additive (Geary, 2013), negative interaction, depotentiation, and even negative synergy. Synergy is most often defined in relation to the realms of pharmacology and medicine. This is because many diseases require treatment that consists of 'cocktails', or mixtures of various drugs taken at once. This is particularly true in cancer treatment. This potentially allows for maximizing the therapeutic effect while minimizing the adverse effects, or side-effects, upon being treated with a given drug regimen (Greco et al., 1995; Greco et al., 1996; Foucquier and Guedj, 2015). If two drugs act synergistically, lower doses of each drug could potentially be used which could allow for less adverse effects while still providing the desired outcome, such as cancer cell death (Roell e, 2017).

METABOLISM(S) OF ENVIRONMENTAL XENOBIOTICS

Organisms are frequently exposed to environmental factors including radiations, metals, pesticides, and other foreign toxic chemical compounds, which could be both the naturally occurring bioactive compounds of plant foods, synthetic chemical agents in food additives, medicines, and environmental pollutants. These factors enter into the organism's body through ingestion, such as food materials (Satarug et al., 2003; Tchounwou et al., 2012; Jaishankar t, 2014), inhalation and dermal contact, such as emissions of waste material in the form of smoke, dust particles, fume of chemicals from several industrial activities such as mining, and manufacturing of batteries (Michael et al., 2007). External exposure to the radioactive source including X-rays and gamma rays pass through the body, depositing energy as they go. Internal exposure to radioactive material gets inside the body by eating, drinking, breathing, or injection (from certain medical procedures). Radionuclides may pose a serious health threat if significant quantities are inhaled or ingested. The rate at which the body metabolizes and eliminates the radionuclide following ingestion or inhalation has not been well studied. It has been reported that heavy metals and pesticides mainly enter into the animal body through exposure from agricultural practices, working and smoking in pesticide- and heavy metal-infested environments. Household practices are major contributors in this context (International Agency for Research on Cancer (IARC) 1993; Singh et al., 2017).

Almost all organs of our body contain xenobiotics biotransformation systems but major part of toxicants is metabolized in the liver. The biotransformation of xenobiotics by organs of our body is carried out to convert the highly toxic chemicals into relatively less toxic molecules through a series of biochemical reactions, also called metabolism of xenobiotics. Biotransformation of xenobiotics refers to an attempt to detoxify the hazardous chemicals by any organism. But it is not always the same. In some cases, such xenobiotic agents after being metabolized become more toxic than their original compound, which could be a lethal metabolite. This process is called bioactivation. For instance, many of the organophosphorous insecticide compounds after metabolism produce more toxic metabolites than their parent compounds such as conversion of parathion to paroxan.

Xenobiotics metabolism takes place in two different phases. Phase I of xenobiotics metabolism includes redox reactions (oxidation and reduction), peroxygenation, hydrolysis, transamination, isomerization, dehalogenation, and epoxidation. These biochemical reactions of biotransformation of xeno-biotics are chiefly catalyzed by the hepatic enzymes (**monooxygenases** or cytochromes P450). These Phase I reactions convert exogenous compounds to their derivatives which become relatively more hydrophilic in nature. These molecules then enter into Phase II reactions. However, products produced by Phase I reactions may as such get excreted, if the polar solubility of products permits translocation.

Phase II biotransformation reactions principally involve either conjugation or synthesis reactions. Those compounds which do not get converted into relatively more hydrophilic forms by means of Phase I reactions are allowed to undergo Phase II reactions. Common conjugation reactions of such xeno-biotics include acetylation, glucuronidation, and combinations with glycine. Metabolism of any of the xenobiotic agents seldom follows a single pathway. Usually, a fraction is excreted unchanged, and the rest is excreted or accumulated as their metabolites. Significant differences in metabolic mechanisms do exist between different species of animals. For instance, cats are deficient in different forms of the enzyme glucuronyl transferase. Hence their ability to conjugate different xenobiotic compounds such as morphine and phenols is compromised. The increased tolerance to subsequent exposures of a toxicant, in some instances, is due to enzyme induction initiated by the previous exposure (Singh et al., 2017). A summary of Phase I and Phase II reactions is displayed in Figure 5.1.

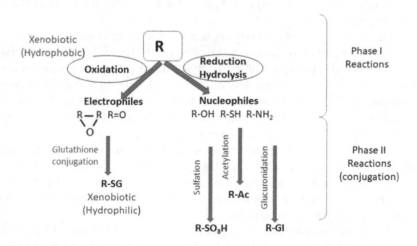

FIGURE 5.1.   Xenobiotics metabolism

SYNERGISTIC EFFECTS

Living organisms in nature are frequently exposed to a mixture of several anthropogenic chemical compounds (heavy metals, pesticides, and toxic gases) simultaneously. These substances have been reported to cause toxicity in animals and in the key organs of the humans (Omiecinski c, 2011; Oesch et al., 2014). Therefore, the combined interactions between chemical substances of different nature and the animal systems are very important (Oesch et al., 2014). The human genome is subjected to constant endogenous as well as exogenous assaults. To maintain genomic integrity, cells have evolved a highly complex signal transduction network to recognize, signal, and repair DNA lesions. The chief components of DNA damage response system are PI3-kinase related protein kinases (PIKKs), ataxia telangiectasia mutated (ATM), DNA-dependent protein kinase catalytic subunit, also known as DNA-PKcs, and ATM and RAD3 related (ATR). These enzymes orchestrate hierarchical pathways that protect genomic integrity, which is called DNA damage response (DDR). RPA-coated single-stranded DNA activates ATR and DSBs, which are particularly hazardous to the cells repaired by ATM and DNA PK. When DSBs occur, the breaks are recognized by the MRN complex (MRE11A-Rad50-NBS1) and then ATM is recruited to the damage site (Shao, 2018; Menolfi and Zha, 2020).

Constant environmental assaults are inflicted on human T cells inducing DNA damage. Several environmental factors affect either Rheumatoid arthritis (RA) susceptibility, or the pathophysiology of RA, and several such environmental factors have been identified. Increased caffeine intake, a history of heavy smoking, and exposure to air pollution are all associated with increased rheumatoid factors. Caffeine reportedly uncouples the cell cycle progression from DNA repair via the mechanism of directly repressing the G2/M checkpoint by inhibiting ATM kinase (Ren et al., 2014; Sharif et al., 2017). However, reports regarding the effects of caffeine are controversial. A cohort study comparing the effects of caffeinated, decaffeinated, and total coffee intake in women did not support the distinct association between caffeinated coffee or caffeine intake with RA pathogenesis (Mikuls et al., 2002). Complex environmental carcinogens, such as tobacco smoke, also serve as genotoxic agents that trigger H2AX phosphorylation and DDR signals. A mechanistic study reported that citrullinated antigens, the self-antigen for ACPA autoantibody of RA patients, are formed in individuals with a history of smoking via activation of peptidylarginine deiminase at the sites of inflammation. A subsequent study reported that nicotine in tobacco is a strong inducer of p53 gene mutations (Tanaka et al., 2007).

Nicotine also promotes neutrophil extracellular traps (NETosis) production in the neutrophil of RA patients, which play an important role in RA-associated inflammation. The formation of NETosis needs a systematic process for DNA which is regulated by DDR signals. Those studies suggest that DDR benefits nicotine-regulated NETosis development and the DDR signal way in T cell and neutrophil of RA patients may be different. The DNA damage response (DDR) is a specific and hierarchical network that includes cell cycle checkpoints, DNA repair, and DNA-damage tolerance pathways. Recent studies suggest that this condition is associated with deficits in telomere maintenance and overall genomic instability in the T cells of RA patients (Shao, 2018). DNA damage is considered an important step in the carcinogenic process. Exposure to mutagenic carcinogens such as tobacco-derived aromatic amines (AA), polycyclic aromatic hydrocarbons (PAHs), and their derivatives such as nitro-PAHs, can cause DNA adduct formation after metabolic activation. These carcinogenic DNA adducts occur in the early stage of mutation and, when the damage is excessive and left unrepaired, may ultimately lead to induction of lung cancer. The combined effects of *GSTM1* genotype with *NAT2* are thought to be potential concurrent risk factors of individual genetic susceptibility to DNA damage. Some workers have constructed a model that included all of the possible combinations among *GSTM1* and *NAT2*, with the low-risk combination of *GSTM1* present and *NAT2* rapid acetylator, as a reference group. The high-risk group, defined as the combination of *GSTM1* null plus *NAT2* slow, was highly significantly associated with an increased level of DNA adducts in their lungs as compared to the low-risk combination. In the case of blood analysis, the sample number was small; thus, their combined analysis suffered from a lack of statistical power to find such associations (Lee et al., 2010).

CONCLUSION

Anthropogenic chemicals present in the environment consistently pose myriad challenges to organisms, through induction of toxicity or mutagenesis. Mutations are generated due to alterations in the base sequences of DNA or the addition of adducts with the purines or pyrimidines bases. This chemical modification of these bases prevents the accurate DNA replication and/or transcription into RNA. The mutation of bases of DNA may also take place by the processes of addition or deletion of base(s), tautomerization, and deamination. It has been documented that every day each human cell undergoes nearly 10,000 'insults' but most of these are repaired or corrected by any one of varied DNA repair systems present in the cell. In order to protect the cells from accumulation of these mutations, the cells typically divide (mitosis) 40–70 times (called Hayflick number) prior to entering into the apoptosis i.e. programmed cell death. In fact, the rise in accumulation of mutations in DNA induced by different chemicals present in the environment happens to be the primary cause of cancer, which is signified by uncontrolled/unregulated cell division and growth. Carcinogenic or mutagenic chemicals such as the alkylating agents present in the environment might act synergistically to cause DNA damage; some of them being Agent Orange, dichlorodiphenyltrichloroethane, intercalating agents, plasticizers such as bisphenol A, and the derivatives of bisphenolA, phthalates, and polyvinyl chloride, and triclosan. There are some environmental chemicals which are known to cause DNA damage indirectly through bringing epigenetic changes i.e. via chemical modification of the nucleotide bases without altering any change into the base sequence. This event results in the abnormal or incorrect expression of the genetic information which may be transferred from one generation to another generation and is called transgenerational inheritance (Ackerman and Horton, 2018).

REFERENCES

Ackerman, S., Horton, W. 2018. Effects of environmental factors on DNA: Damage and mutations. *Green Chemistry* 4; 109–128.

Cedergreen, N. 2014. Quantifying synergy: A systematic review of mixture toxicity studies within environmental toxicology. *PLOS ONE* 9(5); e96580.

Chou, T.C. 2010. Drug combination studies and their synergy quantification using the Chou Talalay method. *Cancer Research* 70(2); 440–446.

Foucquier, J., Guedj, M. 2015. Analysis of drug combinations: Current methodological landscape. *Pharmacology Research Perspectives* 3(3); e00149.

Geary, N. 2013. Understanding synergy. *American Joyrnal of Physiology Endocrinology and Metabolism* 304(3); E237–E253.

Greco, W.R., Bravo, G., Parsons, J.C. 1995. The search for synergy: A critical review from a response surface perspective. *Pharmacology Review* 47(2); 331–385.

Greco, W.R., H Faessel, H., Levasseur, L. 1996. The search for cytotoxic synergy between anticancer agents: a case of Dorothy and the ruby slippers? *Journal of the National Cancer Institute* 88(11);699–700. doi: 10.1093/jnci/88.11.699.

International Agency for Research on Cancer (IARC) 1993. IARC monographs on the evaluation of carcinogenic risks to humans; volumes 1-58. 1993, Lyon: IARC, 1972–1993.

Jaishankar, M., Tseten, T., Anbalagan, N., Mathew, B.B., Beeregowda, K.N. 2014. Toxicity, mechanism and health effects of some heavy metals. *Interdisciplinary Toxicology* 7(2); 60–72.

Lee, M.-S., Su, L., Christiani, D.C. 2010. Synergistic effects of *NAT2* slow and *GSTM1* null genotypes on Carcinogen DNA damage in the lung. *Cancer Epidemiology Biomarkers and Prevention* 19(6); 1492–1497.

Menolfi, D., Zha, S. 2020. ATM 2020, ATR and DNA-PKcs kinases—the lessons from the mouse models: inhibition ≠ deletion. *Cell Bioscience* 10; 8.1–15. https://doi .org/10.1186/s13578-020-0376-x

Michael, S., et al. 2007. Medical Toxicology and Public Health--Update on Research and Activities at the Centers for Disease Control and Prevention, and the Agency for Toxic Substances and Disease Registry. *Journal of Medical Toxicology*, 3(3); 139+.

Mikuls, T.R., Cerhan, J.R., Criswell, L.A., Merlino, L., Mudano, A.S., Burma, M., Folsom, A.R., Saag, K.G., Burma, M. 2002. Coffee, tea, and caffeine consumption and risk of rheumatoid arthritis: Results from the Iowa Women's Health Study. *Arthritis Rheumatology* 46(1); 83–91.

Oesch, F., Fabia, E., Guth, K., Landsiedel, R. 2014. Xenobiotic-metabolizing enzymes in the skin of rat, mouse, pig, guinea pig, man, and in human skin models. *Archives of Toxicology* 88(12); 2135–2190.

Omiecinski, C.J., Vanden, H.J.P., Perdew, G.H., Peters, J.M. 2011. Xenobiotic metabolism, disposition, and regulation by receptors: From biochemical phenomenon to predictors of major toxicities. *Toxicology Sciences* 120(Supplement 1); S49–S75.

Ren, L., Tan, X.J., Xiong, Y.F., Xu, K., Zhou, Y., Zhong, H., Liu, Y., Hong, Y.H., Liu, S.J. 2014. Transcriptome analysis reveals positive selection on the divergent between topmouth culter and zebrafish. *Gene* 552(2); 265–271.

Roell, K.R., Reif, D.M., Motsinger-Reif, A.A. 2017. An introduction to terminology and methodology of chemical synergy-perspectives from across disciplines. *Frontiers in Pharmacology* 8; 1–11.

Satarug, S., Baker, J.R., Urbenjapol, S., Haswell-Elkins, M., Reilly, P.E., Williams, D.J., Moore, M.R. 2003. A global perspective on cadmium pollution and toxicity in non-occupationally exposed population. *Toxicology Letters* 137(1–2); 65–83.

Shao, L. 2018. DNA damage response signals transduce stress from rheumatoid arthritis risk factors into T cell dysfunction. *Frontiers in Immunology* 9; 3055.

Sharif, K., Watad, A., Bragazzi, N.L., Adawi, M., Amital, H., Shoenfeld, Y. 2017. Coffee and autoimmunity: More than a mere hot beverage! *Autoimmunity Reviews* 16(7); 712–721.

Singh, N., Gupta, V.K., Kumar, A., Sharma, B. 2017. Synergistic effects of heavy metals and pesticides in living systems. *Frontiers in Chemistry* 5; 1–9.

Tallarida, R.J. 2001. Drug synergism: Its detection and applications. *Journal of Pharmacology Experimental Therapeutics* 298(3); 865–872.

Tanaka, T., Huang, X., Halicka, H.D., Zhao, H., Traganos, F., Albino, A.P., Dai, W., Darzynkiewicz, Z. 2007. Cytometry of ATM activation and histone H2AX phosphorylation to estimate extent of DNA damage induced by exogenous agents. *Cytometry A* 71(9); 648–661.

Tchounwou, P.B., Yedjou, C.G., Patlolla, A.K., Sutton, D.J. 2012. Heavy metal toxicity and the environment. *Experientia Supplementum* 101; 133–164.

Introduction to phytochemicals and their actions on DNA. A. DNA-damaging action. B. DNA-protective function

6

INTRODUCTION TO PHYTOCHEMICALS

Phytochemicals are an important part of traditional medicine and have been explored in detail for possible inclusion in modern medicine as well. Often, these phytochemical compounds serve as the backbone in the synthesis of novel therapeutic molecules. Plants are used in the synthesis of numerous medicinal compounds, whose characterization has led to the discovery of new, low cost drugs with high therapeutic potential (Kumar and, Khanum, 2012; Ukwuani et al., 2013). The exploitation of plants in medicine is very old, for instance, in 28 AD the Greek physician Dioscorides wrote *De Materia Medica*, which included 600 medical plants. It was the leading text on pharmacology until the renaissance. Hippocrates might have prescribed willow tree leaves to abate fever. Salicin having anti-inflammatory and pain-relieving properties was originally extracted from the white willow tree and later synthetically produced to become the staple over-the-counter drug (Huie, 2002; Njerua et al., 2013).

Medicinal plants contribute significantly to rural livelihoods. Apart from the traditional healers practicing herbal medicine, many people are involved in collecting and trading medicinal plants. The result is an increased demand in both local and international markets as well as bio-prospecting activities searching for sources of new drugs. The World Health Organization (WHO) estimates that 80% of the world's population depends on medicinal plants for their primary health care (Akerele, 1993; Mothana et al., 2010; Gupta et al. 2010; Ngoci et al., 2011; Prakash and Sandhu, 2012). Drugs derived from unmodified natural products or semisynthetic drugs obtained from natural sources accounted for 78% of the new drugs approved by United States Food and Drug Administration (FDA) between 1983 and 1994 (Suffredini et al., 2006; Ngoci et al., 2011). A survey on medicinal plant usage by American public showed an increase from 3% in 1993 to over 37% in 1998 (Briskin, 2000). This shift to herbal drugs has been facilitated by factors like the low cost of herbal drugs, endearing them with the poor mass of developing world; the 'green' movement in the developed world that campaigns on the inherent safety and desirability of natural products and the individualistic philosophy of Western society that encourages self-medication, with many people preferring to treat themselves with phytomedicines (Sharma et al., 2012; Njerua et al., 2013).

On the other hand, many plants might appear innocent and harmless at first glance, as many of these plants possess properties to produce toxins dangerous to humans and other organisms. Phytotoxins are toxins or secondary metabolites produced by different plant species with adverse effects on the health of animals and humans. The most important effect of phytotoxins is to inhibit the physiological and biochemical processes in plants as well as animals (Möbius and Hertweck, 2009). These plants cause harm to the organisms either via indirect and/or direct exposure to phytochemicals. The indirect effect is mediated through release of phytochemicals into the environment by leaching, root exudation, and volatilization, and passively through decomposition of plant residues, which may be further taken up by organisms via inhalation, dermal contact, and ingestion. It has been observed to interrupt the growth of organisms by interrupting their physiological and biochemical processes (Bender et al., 1998). In case of direct exposure, the plant molecules or plant parts are directly taken up by any organism either alone or in combination with other food materials (Singh and Sharma, 2018).

Bio-activity of natural plant products is due to the presence of various phytochemicals, generally explained for the defense mechanisms of plants against attack by different pathogens, herbivory, inter-plant competition, and against abiotic stresses (Briskin, 2000; Ruba et al., 2013). These phytochemicals inadvertently protect humans against pathogens as anti-microbial medicines. Some phytochemicals are known to have therapeutic and prophylactic properties; they provide nutrition for normal cell health and repairs, inhibit carcinogens, and act as antioxidants (Ogunwenmo et al., 2007; Ngoci et al., 2011). Phytochemicals exert their medicinal effects by acting synergistically or additively and this eliminates the risks of any side-effects associated with the predominance of a single xenobiotic compound, give the herbal drug(s) a broad spectrum of activity, as well as decreasing the chances of the pathogens developing resistance or adaptive responses (Briskin, 2000; Olila, et al., 2001).

PHYTOCHEMICALS

Secondary metabolites synthesized by different plant species are considered as phytochemicals. These compounds can have complementary and/or overlapping mechanisms of action in the body, including antioxidant effects, modulation of enzyme actions, stimulation of the immune system, modulation of hormone metabolism, antibacterial and antiviral effects, interference with DNA replication and physical action whereby some may bind physically to cell walls thereby preventing the adhesion of pathogens to human cell walls (Ngoci, et al., 2011). Some of these phytochemicals are discussed in the following sections.

ALKALOIDS

An alkaloid is a plant-derived compound that is toxic or physiologically active chemical agent. Some alkaloids such as isopteropodine and pteropopine have anti-microbial activity whereby they act by promoting white blood cells to dispose harmful microorganisms and the cell debris (Ogunwenmo et al., 2007). Highly aromatic planar quaternary alkaloids such as berberine, piperine, sanguinarine, and harmane work by intercalating the DNA and cell wall (Cowan, 1999). Others may act by simulating neurotransmitters such as acetylcholine, dopamine, and serotonin, and may influence the central nervous system (CNS) at the synapses. However, some types of alkaloids are hallucinative, addictive, and toxic and hence used as arrow poison for hunting wild game (Ogunwenmo et al., 2007; Ngoci et al., 2011).

TANNINS

Tannins are astringent, bitter plant polyphenols that either bind with and precipitate or shrink the proteins. They have physiological role by acting as antioxidants through free radical scavenging activity, chelation of transition metals, inhibition of prooxidative enzymes and lipid peroxidation (Navarro et al., 2003; Vit et al., 2008; Ngoci et al., 2011), hence modulating oxidative stress and preventing degenerative diseases. Also, they inhibit tumor growth by inducing apoptosis (Scalbert et al., 2005) and inhibiting mutagenecity of carcinogens (Okuda, 2005; Ngoci, et al., 2011). They exhibit anti-microbial activity by complexing with the nucleophilic proteins by establishing hydrogen bond interactions. In addition, some saponins like Radix Notoginseng have been described to increase the blood flow of the coronary arteries, to prevent platelet aggregation and to decrease the consumption of oxygen by heart muscles (Dong et al., 2005).

PHYTOSTEROIDS

Phytosteroids are plant steroids that may or may not act as weak hormones in the body. They share a common basic ring structure with animal steroids though they are not equivalent because of varying chemical groups attached to the main ring at different positions (Ngoci, et al., 2011). They are mainly used to treat reproductive complications such as treatment of venereal diseases, used during pregnancy to ensure an easy delivery, as well as to promote fertility in women and libido in men. They also act as sex hormones derivatives; for example, they can be metabolized to either androgen or estrogen-like substances and hence they are potential source of contraceptives (Edeoga et al., 2005; Ngoci et al., 2011). They also act as anti-microbial, analgesic, and anti-inflammatory agents. Phytosteroids are also used in treating stomach ailments and in decreasing serum cholesterol levels (Ngoci, et al., 2011).

TERPENOIDS

These are derivatives of isoprene molecule having a carbon skeleton built from one or more of C15 units (Harborne, 1973). They exert their roles as antibacterial, antifungal, antiviral, antiprotozoan, antiallergen, immune booster, and antineoplastic agents (Roberts, 2007; Ngoci, et al., 2011). Petalostemumal has been demonstrated to exhibit activity against *B. subtilis*, *S. aureus*, and *C. albicans* and to a lesser extent to Gram-negative bacteria (Cowan, 1999; Ngoci, et al., 2011). This is related to physicochemical characteristics of the active principle (such as lipophilicity and water solubility), lipid composition and net surface charge of the bacterial membranes. These phytochemicals can cross the cell membranes, penetrating the interior of the cell and interacting with intracellular targets critical for antibacterial activity (Trombetta et al., 2005).

They are also used to alleviate epilepsy, to get relief from cold, influenza, cough, and acute bronchial disease (Ngoci, et al., 2011).

CARDIAC GLYCOSIDES

Cardiac glycosides (also called cardenoloids) occur as a complex mixture together in the same plant and most of them are toxic. However, many of them have pharmacological activity especially in case of heart ailments (Harborne, 1973). They are used in treatment of congestive heart failure, whereby they inhibit Na+/K+-ATPase pump that causes positive ionotropic effects and electrophysiological changes. This strengthens the heart muscle and the power of systolic contraction against congestive heart failure (Ogunwenmo et al., 2007; Ngoci et al., 2011). They are also used in treatment of atrial fibrillation and flutter. The cardiac glycosides also function as emetics and as diuretics (Harborne, 1973; Ngoci et al., 2011).

SOURCE OF DIFFERENT PHYTOCHEMICALS

Antioxidant phytochemicals exist widely in cereal grains, fruits, vegetables, edible microalgae, macrofungi, and medicinal plants. Common fruits including berries, Chinese date, Chinese wampee, grape, guava, persimmon, pomegranate, plum, and sweetsop are rich in antioxidants. Additionally, wild fruits including the fruits of *Caryota mitis, Eucalyptus robusta, Eurya nitida, Gordonia axillar, Lagerstroemia indica, Lagerstroemia speciose Melastoma sanguineum,* and *Melaleuca leucadendron* also have high antioxidant potential and total polyphenolic contents. Instead, the wastes from fruits such as its peel and seeds are also rich in antioxidant phytochemicals, including kaempferol, catechin, epicatechin, gallic acid, chlorogenic acid, and cyanidin 3-glucoside. Some vegetables including loosestrife, cowpea, caraway, penile leaf, lotus root, broccoli, Chinese toon bud, sweet potato leaf, pepper leaf, soy bean (green), ginseng leaf, and chives are recorded to contain high antioxidant and total phenolic compounds. Among cereal grains pigmented rice, such as black rice, red rice, and purple rice, possess high contents of antioxidant phytochemicals (flavones and tannins). Among selected Chinese medicinal plants, the highest antioxidant capacities and phenolic contents are found in *Dioscorea bulbifera, Eriobotrya japonica, Tussilago farfara* and *Ephedra sinica*, and several flowers including edible and wild ones also have high contents of antioxidant phytochemicals (Zhang et al., 2015).

PHYTOCHEMICALS AND THEIR MODES OF ACTION

Since ancient times, plants and their by-products have been utilized in treatment of many severe health abnormalities. These medicinal plants significantly contribute to the livelihoods of people mostly in rural areas. Apart from their traditional use as healers, several people are involved in development of modern drugs with better healing capacity using these plants. The antioxidative potential of any plant mainly depends on its polyphenol and carotenoid contents. They contribute the most to the antioxidant properties of foods/plants. For example, beta-carotene, myricetin, quercetin, and kaempferol are the main antioxidant phytochemicals found in Cape gooseberry, and anthocyanins and ellagitannins are the major antioxidant compounds among the phytochemicals of strawberry. In addition, flavonoids isolated from *Euterpe oleracea* pulp present an important antioxidant activity measured by oxygen radical absorbance capacity (Petrovska, 2012). Natural polyphenols are the most abundant antioxidants in human diets, and their radical scavenging activities are related to substitution of hydroxyl groups in the aromatic rings of phenolics. The plant variety, geographic region, growing season, and storage can all influence the concentrations of polyphenols in food. Dietary polyphenols could be classified into five classes: flavonoids, phenolic acids, stilbenes, tannins, and coumarins. Flavonoids can be further categorized as flavonols, flavones, flavanols, flavanones, anthocyanidins, and isoflavonoids. Total phenolic content and total antioxidant activity in phytochemical extracts of different fruits may have a direct relationship. When the fruits contain higher total phenolic contents, they possess stronger antioxidant activity. For example, the scavenging activity of grape seed extract against ABTS radical was strongly linked with the level of phenolic compounds. Carotenoids are a group of phytochemicals that are responsible for the yellow, orange, and red colors of the foods. Beta-carotene, lycopene, lutein, and cryptoxanthin are the main carotenoids in diet, and fruits and vegetables are the major sources of carotenoids in human diet. For example, tomato is rich in lycopene, which is also responsible for its characteristic red color (Pandey and Rizvi, 2009).

In a study, the cytoprotective activities of ethanolic, ethyl acetate, chloroform, and aqueous extracts of *C. papaya* have been performed. The oxidation of erythrocytes is a good model for the oxidation of bio membranes in general. From the *in vitro* Erythrocyte Cytoprotection Assay it was observed that Co-treatment of erythrocytes with ferrous sulfate and extracts showed that CP leaves extract had the ability to restore the crenated RBCs to their native biconcave form, while pretreatment with extracts followed by treatment with oxidant showed delayed damage. These findings thus demonstrated that the *C.papaya* leaf extracts possess significant cytoprotective property (Kadri et al., 2016).

DETRIMENTAL ROLE OF PHYTOCHEMICALS AND THEIR MODES OF ACTION

Although many studies have found protective roles for antioxidant phytochemicals in chronic diseases, other studies found some discrepancies. For example, according to one study, fruit juices from red grape, strawberry, cherry, or sour cherry showed very strong free radical scavenging activity in the 2,2-diphenyl-1-picrylhydrazyl radical scavenging capacity assay and the beta-carotene bleaching assay, but they did not show cytotoxic effects on HT29 cells using the same concentrations. That is, there was no correlation between the antioxidant activity and antiproliferative effects of these fruit juices in the HT29 cells. In another study, effects of several antioxidant phytochemicals on the tumor-promoting activity of 3,31,4,41-tetrachlorobiphenyl were examined *in vivo*. Coenzyme 10 increased cell proliferation in normal hepatocytes, whereas the other antioxidants such as ellagic acid, beta-carotene, curcumin, N-acetyl cysteine, resveratrol, lycopene, and a tea extract had no effect in either normal or PGST-positive hepatocytes. The results showed that none of the antioxidant phytochemicals produced a clear decrease in the tumor-promoting activity of 3,31,4,41-tetrachlorobiphenyl in rats. In addition, the workers observed by a systematic evaluation that the concentrations of vitamin C, vitamin E, and beta-carotene in dietary were inversely associated with gastric cancer risk, while no such association was observed for blood levels of these antioxidants. Although some epidemiological studies showed a relationship between the low incidence of cancer and the intake of plant-based foods, at present there is no conclusive proof that high antioxidant activity would result in high anticancer activity (Zhang, 2015; Wongnarat and Srihanam, 2017; Velić et al., 2018).

Some studies have also indicated that supplemental antioxidants cannot decrease the risks for some diseases and could even play an inverse role. To assess the effect of antioxidant supplements on mortality in randomized primary and secondary prevention trials, the effect of antioxidant supplements on all-cause mortality was analyzed with random-effects meta-analyses and reported as relative risk with 95% confidence intervals. Meta-regression was used to assess the effect of covariates across the trials. The results showed that vitamin C and selenium had no significant effect on mortality; treatment with beta-carotene, vitamin A, and vitamin E may increase mortality. In addition, a review paper also pointed out that the majority of the studies did not support a possible role of antioxidant supplementation in reducing the risk of cardiovascular disease. Thus, the application of antioxidant therapy in certain diseases needs long-term investigations on large-scale cohorts (Hajhashemi et al., 2010).

Among all phytochemicals, alkaloids are considered as a group of naturally occurring chemical compounds that are mostly responsible for toxicological consequences of a particular plant. These groups contain basic nitrogen atoms and some related chemical compounds also include weakly acidic and neutral properties. A number of pharmacological activities are associated with alkaloids such as antimalarial (quinine), anticancer (homoharringtonine), antiasthma (ephedrine), antihyperglycemic activities (vincristine and vinblastine) cholinomimetic (galantamine), antiarrhythmic (quinidine), analgesic (morphine), vasodilatory (vincamine), and antibacterial (chelerythrine). The adverse effects of alkaloids have been documented by several researchers (Matsuura and Fett-Neto, 2015; Adibah and Azzreena, 2019). For instance, berberine, an isoquinoline alkaloid, present in different parts such as roots, stem, bark, rhizome, fruit, and leaves (rarely) of several plant species (mostly in barberry) contains toxic chemicals.

The potential antitumor activity of berberine hydrochloride has always been a subject of considerable interest because of the known capability of berberine to bind with nucleic acids. Its ability to bind specifically to oligonucleotides and to stabilize DNA triplexes or G-quadruplexes via telomerase and topoisomerase inhibition accounts for its antiproliferative activity (Tan et al., 2011; Hao et al., 2014). In addition, berberine is reported to induce a significant hormetic dose response, in which the low dose of berberine strongly stimulates the growth of cancer cells, while at high doses it acts as anticancer agents (Bao et al., 2015; Singh and Sharma, 2018). Moreover, its extensive occurrence in various plant species and low toxicity suggest that berberine hydrochloride has the potential to become an effective antitumor

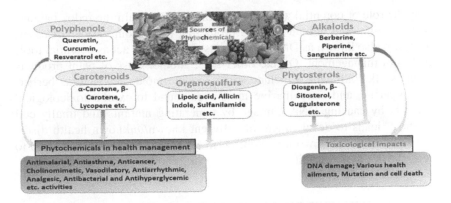

FIGURE 6.1. Multiplex functions of varied plant-based molecules.

agent in future. In addition to berberine, the cytotoxic and cytostatic effects of sanguinarine (an alkaloid) on a variety of human cancer cells, including human epidermoid carcinoma, erythroleukemia, prostate cancer, pancreatic carcinoma, colon cancer, breast cancer, lung cancer, promyelocytic leukemia, and bone cancer, have been reported. Sanguinarine exhibits the highest cytotoxicity among benzophenanthridine alkaloids (Singh and Sharma, 2018). Sanguinarine, a phytotoxin, is known to kill the animal cells by inhibiting the function of the Na+-K+-ATPase, a transmembrane protein. The enzyme, Na+-K+-ATPase, is responsible for maintaining the resting potential across the cell membrane by regulating the volume of a cellula through pumping sodium out of the cells and potassium into the cells. This phenomenon takes place against the gradients of Na^+ and K^+ concentrations. The pumping event of Na^+/K^+ is an energy (ATP) dependent process as it is mediated via active transport system. This is crucial for regulating cellular physiology. The Na^+/K^+ ATPase is also known to act as a signal transducer in order to regulate the activity of mitogen-activating protein kinases (MAPK) pathway, the concentration of reactive oxygen species (ROS), and the level of intracellular calcium. The varied functions of phytochemicals are summarized in Figure 6.1.

CONCLUSION

Plant products are known to perform both as medicines as well as toxicants but both of these kinds of phytochemicals exhibit significant biomedical implications. The polyphenols, carotenoids, organosufurs, and phytosterols play the role of potential antioxidants and help maintain the redox balance in the animal and human systems protecting them from various diseases. They are also known as DNA protectants. Hence, they act as potential tools in health management. On the other side, another class of plant products such as alkaloids which include some specific molecules such as berberine, piperine, and sanguinarine have been found to exhibit toxicological impacts by causing DNA damage by generating mutants and finally cell death. It leads to development of different known/unknown health disorders. However, lot of work is required to be carried out in this direction to reap the optimum benefit.

REFERENCES

Adibah, K.Z.M., Azzreena, M.A. 2019. Plant toxins: Alkaloids and their toxicities. *GSC Biological and Pharmaceutical Sciences* 6; 21–29.

Akerele, O. 1993. Nature's medicinal bounty: Don't throw it away. *Traditional Medicine* 14; 390–395.

Bao, J., Huang, B., Zou, L., Chen, S., Zhang, C., Zhang, Y., Chen, M., Wan, J.B., Su, H., Wang, Y., He, C. 2015. Hormetic effect of berberine attenuates the anticancer activity of chemotherapeutic agents. *PLoS One* 10; e0139298.

Bender, D.J., Contreras, T.A., Fahrig, L. 1998. Habitat loss and population decline: A meta-analysis of the patch size effect. *Ecology* 79; 517–533.

Briskin, D.P. 2000. Medicinal plants and phytomedicines. Linking plant biochemistry and physiology to human health. *Journal of Plant Physiology* 124; 507–514.

Cowan, M.M., 1999. Plant products as anti-microbial agents. *Clinical Microbiology Reviews* 12; 564–582.

Dong, T.T.X., Zhao, K.J., Huang, W.Z., Leung, K.W., Tsim, K.W.K. 2005. Orthogonal array design in optimizing the extraction efficiency of active constituents from roots of panax notoginseng. *Phytotherapy Research* 19; 684–688.

Edeoga, H.O., Okwu, D.E., Mbaebie, B.O., 2005. Phytochemical constituents of some Nigerian medicinal plants. *African Journal of Biotechnology*, 4; 685–688.

Gupta, V.K., Shukla, C., Bisht, G.R.S., Kumar, S., Thakur, R.L., 2010. Detection of anti-tuberculosis activity in some folklore plants by radiometric BACTEC assay. *Letters in Applied Microbiology* 52; 33–40.

Hajhashemi, V., Vaseghi, G., Pourfarzam, M., Abdollahi, A. 2010. Are antioxidants helpful for disease prevention? *Research in Pharmaceutical Science* 5; 1–8.

Hao, Z. et al. 2014. Global integrated drought monitoring and prediction system. *Science Data* 1:140001 doi:10.1038/sdata.2014.1.

Harborne, J.B., 1973. *Phytochemical Methods*. London: Chapman and Hall, 52–114.

Huie, C.W. 2002. Review: Modern sample preparation techniques for the extraction and analysis of medicinal plants. *Analytical and Bioanalytical Chemistry* 373; 23–30.

Kadri, S.U.T., Nikhitha, M., Shlini, P. 2016. Cytoprotective and DNA protective activity of Carica Papaya Leaf extracts. *International Journal of Pharmaceutical Science Invention* 5; 35–40.

Kumar, G.P., Khanum, F. 2012. Neuroprotective potential of phytochemicals. *Pharmacognsy Reviews* 6; 81–90.

Matsuura, H., Fett-Neto, A.G. 2015. Plant alkaloids: Main features, toxicity, and mechanisms of action. *Plant Toxins* 2; 1–15. doi:10.1007/978-94-007-6728-7_2-1

Möbius, N., Hertweck, C. 2009. Fungal phytotoxins as mediators of virulence. *Current Opinion in Plant Biology* 12; 390–398.

Mothana, R.A., Abdo, S.A., Hasson, S., Althawab, F.M., Alaghbari, S.A., Lindequist, U. 2010. Antimicrobial, antioxidant and cytotoxic activities and phytochemical screening of some yemeni medicinal plants. *Evidence-Based Complementary and Alternative Medicine* 7; 323–330.

Navarro, M.C., Montilla, M.P., Cabo, M.M., Galisteo, M., Caceres, A., Morales, C., Berger, I. 2003. Antibaterial, antiprotozoal and antioxidant activity of five plants used in izabal for infectious diseases. *Phytotherapy Research* 17; 325–329.

Ngoci, S.N., Mwendia, C.M., Mwaniki, C.G, Baker, F. 2011. Phytochemical and cytotoxicity testing of Indigofera lupatana. *Journal of Animal & Plant Sciences* 11; 1364–1373.

Njerua, S.N., Matasyoh, J., Mwaniki, C.G., Mwendia, C.M., Kobia, G.K. 2013. A review of some phytochemicals commonly found in medicinal plants. *International Journal of Medicinal Plants Photon* 105; 135–140.

Ogunwenmo, K.O., Idowu, O.A., Innocent, C., Esan, E.B., Oyelana, O.A., 2007. Cultivars of codiaeum variegatum (L.) Blume (Euphorbiaceae) show variability in phytochemical and cytological characteristics. *Journal of Biotechnology* 20, 2400–2405.

Okuda, T. 2005. Systematics and health effects of chemically distinct tannins in medicinal plants. *Journal of Phytochemistry* 66, 2012–2031.

Olila, D., Olwa-Odyek, Opuda-Asibo, J. 2001. Anti-bacterial and antifungal activities of extracts of zanthoxylum chalybeum and warburgia ugandensis, ugandan medicinal plants. *African Health Sciences* 1; 66–72.

Pandey, K.B., Rizvi, S.I. 2009. Plant polyphenols as dietary antioxidants in human health and disease. *Oxidative Medicine and Cellular Longevity* 2; 270–278.

Petrovska, B.B. 2012. Historical review of medicinal plants' usage. *Pharmacognosy Review* 6; 1–5.

Prakash, V., Sandhu, P., 2012. In vitro antimycobiotic and antibacterial action of seed extract of celastrus paniculatus Willd. (Jyotismati). *Journal of Antimicrobials. Photon* 127; 123–132.

Roberts, S.C., 2007. Production and engineering of terpenoids in plant cell culture. *Nature Chemical Biology* 3; 387–395.

Ruba, A.A., Nishanthini, A., Mohan, V.R. 2013. In vitro antioxidant and free radical scavenging activities of leaf of arthocnemum fruticosum Moq (Chenopodiaceae). *The Journal of Free Radicals and Antioxidants Photon* 139; 166–174.

Scalbert, A., Johnson, I.T., Saltmarsh, M. 2005. Polyphenols: Antioxidants and beyond. *American Journal of Clinical Nutrition* 81; 215–217.

Sharma, A., Meena, A., Meena, R., Kumar, A. 2012. Isolation of phytosterols from static culture of Ocimum tenuiflorum L. *The Journal of Bioprocess Technology. Photon* 96; 125–129.

Singh, N., Sharma, B. 2018. Toxicological effects of berberine and sanguinarine. *Frontiers in Molecular Bioscience* 5; 1–7.

Suffredini, B.I., Paciencia, M.L.B., Nepomuceno, D.C., Younes, R.N., Varella, A.D., 2006. Anti-bacterial and cytotoxic activity of Brazilian plant extracts-Clusiaceae. *Memórias do Instituto Oswaldo Cruz* 101; 1590–1598.

Tan, M., Luo, H., Lee, S., Jin, F., Yang, J.S., Montellier, E., Buchou, T., Cheng, Z., Rousseaux, S., Rajagopal, N., Lu, Z., Ye, Z., Zhu, Q., Wysocka, J., Ye, Y., Khochbin, S., Ren, B., Zhao, Y. 2011. Identification of 67 histone marks and histone lysine crotonylation as a new type of histone modification. *Cell* 146; 1016–1028.

Trombetta, D., Castelli, F., Sarpietro, M., Venuti, V., Cristani, M., Daniele, C., Saija, A., Mazzanti, G., Bisignano, G. 2005. Mechanisms of anti-bacterial action of three monoterpenes. *Antimicrobial Agents and Chemotherapy* 49; 2474–2478.

Ukwuani, A.N., Abubakar, M.G., Hassan, S.W., Agaie, B.M. 2013. Antinociceptive activity of hydromethanolic extract of some medicinal plants in mice. *International Journal of Pharmacy Photon* 104; 120–125.

Velić, D., Klarić, A.D., Velić, N., Klarić, I., Tominac, V.P., Mornar, A. 2018. *Chemical Constituents of Fruit Wines as Descriptors of Their Nutritional, Sensorial and Health-Related Properties.* doi:10.5772/intechopen.78796

Vit, K., Katerina, K., Zuzana, R., Kamil, K., Daniel, J., Ludek, J., Lubomir, O. 2008. Mini review: Condensed and hydrolysable tannins as antioxidants influencing the health. *Journal of Medicinal Chemistry* 8; 436–447.

Wongnarat, C., Srihanam, P. 2017. Phytochemical and antioxidant activity in seeds and pulp of grape cultivated in Thailand. *Oriental Journal of Chemistry* 33; 1–9.

Zhang, Y.J., Gan, R.Y., Li, S., Zhou, Y., Li, A.N., Xu, D.P., Li, H.B. 2015. Antioxidant phytochemicals for the prevention and treatment of chronic diseases. *Molecules* 20; 21138–21156.

Phytochemicals as DNA protectants and genotoxicants A. Phytochemicals as genotoxicants. B. Plant extract as DNA protectants. C. Isolated compounds as DNA protectants

7

INTRODUCTION

Since ancient times medicinal plants have been used in the treatment of various diseases. In recent years, awareness regarding application of plant products as natural therapeutics has increased several fold because of numerous side-effects of man-made synthetic drugs. The plant-based drugs are cost effective, safe, and rarely have any side-effects (Kumar et al., 2014). The application of synthetic medicines has been found to undergo slow biotransformation processes, as a result of which they get accumulated in the fat depots of the body. Once their concentration reaches beyond the critical level (the level below that is harmless), they generate adverse effects in the body through over-production of oxidants (ROS and RNS). These free radical species are responsible for the pathogenesis of most of the diseases. The scavenging of these oxidants is thought to be an effective measure to reduce the level of oxidative stress (OS) into the organisms. Antioxidant phytochemicals can be found in many plant sources including food and medicinal plants, and they play an important role in prevention and treatment of chronic diseases caused by OS. They often possess plenty of strong antioxidants with free radicals-scavenging abilities which are also the bases of their bioactivities and health benefits (Zhang et al., 2015).

Naturally occurring compounds are considered to be the most useful agents for cancer prevention and therapy, because of their anticipated multimodal actions and limited toxicity. Phytochemicals may also affect the signaling pathways within the cells, including those regulating cell proliferation, activation of apoptosis, etc. However, the combined regimens of naturally occurring compounds with standard chemotherapeutic drugs are very promising in providing additive or synergistic efficacy. Among many naturally occurring compounds, polyphenols are known to be present in different edible fruits including grapes, berries, pomegranate, walnut, apples, etc. Among these, flavonoids consist of a large group of natural, small molecular weight compounds, ubiquitously present in almost all fruits and vegetables. Polyphenols are reported to have anticarcinogenic and anticancer properties. Numerous studies have been carried out in this context that address the efficiency of polyphenol-containing foods in cancer prevention and therapy. However, there are only limited studies to identify individual components responsible for the anticancer properties of polyphenols. Similarly, little is known about mechanism(s) by which such small molecule inhibitors purified from plants act on tumor cells (Niedzwiecki et al., 2016; Thota et al., 2018).

Some secondary metabolites of plants have been reported to exhibit toxic behavior due to its certain phytochemical ingredients (mainly due to presence of alkaloids), some of them are known to cause DNA damage through generation of free radical species (FRS). For instance, sanguinarine, an alkaloid, has a molecular structure similar to a known carcinogen, which is a polyaromatic hydrocarbon and is a DNA intercalator. Sanguinarine also generates oxidative stress (OS) and adverse impact on endoplasmic reticulum resulting in the formation of unfolded proteins. Sanguinarine is known to induce formation of 8-hydroxyguanine and genetic lesions. However, there are contradictory reports about the *in vitro* and *in vivo* results of sanguinarine regarding its genotoxicity and carcinogenicity issues. Despite this, epidemiological studies have linked the mouthwash reagent that contains sanguinarine with the development of oral leukoplakia. Sanguinarine is also proposed as an aetiological agent in gallbladder carcinoma (Karp et al., 2005; Croaker et al., 2017).

The mechanism involved in genotoxins-induced DNA damage is distortion of the DNA structure via breakage of H-bonds between two complementary base pairs involved in DNA strands stabilization. In order to maintain the genome integrity, it is necessary to repair DNA damage with the help of DNA repair machineries. Any abnormality in DNA repair mechanism can result in genomic instability. Cross-linking agents such as mitomycin C and some aromatic compounds, fungal and bacterial toxins, and metabolic products such as free radicals or reactive oxygen/nitrogen species (ROS/RNS) play crucial role in DNA damage. Free radical-induced DNA damage in living organisms takes place by different mechanisms. These include DNA base and sugar products, single- and double-strand breaks, 8,5'-cyclopurine-2'-deoxynucleosides, tandem lesions, clustered sites, and DNA-protein cross-links. The present chapter illustrates both the beneficial and the harmful effects of plant-based principles. The updates on these points suggest the application of plant products in consultation with concerned physicians or experts only.

A. Phytochemicals as genotoxicants

There are some plant products which are shown to possess DNA-damaging potential. Recent findings suggest an active role of nicotine. One of the major plant products is nicotine, an alkaloid from tobacco present in the tobacco plant that causes carcinogenesis. Nicotine exhibits tumor-promoting potential by inducing DNA damage in different human epithelial and non-epithelial cells. Sanguinarine isolated from a wild plant, *Argemone mexicana*, has been shown to generate chromosomal aberration, micronucleus formation, and DNA damage as shown by comet assay in mouse model *in vivo* system. Sanguinarine is reported to inhibit the activity of epidermal histidase leading to the increase in the levels of keratin formation and tumor promotion (Sarkar, 1948; Becci et al., 1987; Singh and Sharma, 2018a). In another study, the effects of microcystins (MCs)-containing cyanobacteria extract (CE) on damage of DNA in the living organisms was evaluated (Zegura et al., 2011; Shahi et al., 2012). The results suggested significant DNA damage in rice seedlings after exposure to CE. Another studied alkaloid, T-2 Mycotoxin, is a trichothecene mycotoxin. It is a naturally occurring mold by-product of Fusarium spp, a fungus which causes toxicity in humans and animals. Treatment of fasting mice with a single dose of T-2 toxin (1.8 or 2.8 mg/kg body weight) by oral route has been shown to lead to 76% hepatic DNA fragmentation (Shahi et al., 2012).

ISOLATED COMPOUNDS AS GENOTOXICANTS

There is some phytochemical reported to cause adverse effects on genome integrity. For instance, sanguinarine has been reported to inhibit microtubule polymerization and benzophenanthridine cytotoxic activity involves intercalation of double-stranded deoxyribonucleotide (DNA) (Lpus and Panda, 2006; Matkar et al., 2008) and induces fragmentation of DNA. Recent studies stated that the cytotoxicity and DNA-damaging effect of sanguinarine is more specific to cancer cells than to normal cells (Ahmad et al., 2000; Matkar et al., 2008). Usually, isoquinoline alkaloids interact with DNA as intercalators, or they are arranged in a small groove; their external binding with phosphate groups is also possible (Motilal and Gopinatha 2010).

Incubation of HeLa cells with BCL for 4 h showed greater amount of DNA damage (OTM) than 2 h treatment. BCL treatment caused a concentration-dependent rise in the DNA damage in HeLa cells and exposure of HeLa cells with 1 μg/ml BCL caused a10 fold rise in baseline DNA damage, whereas a maximum rise in DNA damage was observed in HeLa cells exposed to 8 μg/ml BCL. DNA-damaging potential of different concentrations of berberine chloride (BCL), an isoquinoline alkaloid has been studied in HeLa cells using alkaline comet assay. The DNA damage has been expressed as olive tail moment (OTM). The results indicated that the cell survival and molecular DNA damage in HeLa cells treated with BCL has an inverse correlation indicating that with increased DNA damage cell survival declined. They have also demonstrated that antineoplastic effect of BCL is mainly due to its ability to cause damage to the cellular genome. The cytotoxic effect of BCL may be due to the induction of free radicals and lipid peroxides that induced DNA damage and loss of repair capacity of HeLa cells leading to cell death. Intercalation of BCL into the DNA, inhibition of topoisomerase II and PARP may have caused molecular DNA damage that would have subsequently led to cell death. Transactivation of NF-κ Band cyclooxygenase II and inhibition of activator protein 1, cyclins, and p53 may have also contributed in their own way to induce cytotoxicity in HeLa cells exposed to BCL (Jagetia and Rao, 2015).

Topotecan is a semisynthetic derivative of camptothecin, a cytotoxic, quinoline-based alkaloid extracted from the Asian tree *Camptotheca acuminate*. Topotecan, a commonly used chemotherapeutic agent and topoisomerase inhibitor, acts against many cancers, mainly cervical cancer and small cell lung cancer, by creating double-stranded DNA damage. Most of these medically exploited alkaloids function as therapeutics by mainly provoking DNA damage, inducing apoptosis, and acting as antiproliferative agents; however, their associated toxicity urges the finding of new natural compounds that have the potential to selectively target cancerous cells.

As such, alkaloids that promote apoptosis via inducing DNA damage seem to be a great option for cancer treatment (Habli et al., 2017). Hirsutine is an indole alkaloid isolated from *Uncaria rhynchophylla* and found in Yokukansan. Hirsutine has been reported to induce DNA damage as manifested by the up-regulation of γH2AX, a marker of DNA breakage, and to increase the expression of p-p38 MAPK. Preliminary results of the anticancer *in vitro* efficacy of hirsutine showed that this alkaloid alone is capable of modulating the survival pathways and causing DNA damage, making it worthy to be incorporated with other drugs for the clinical treatment of TNBC (Habli et al., 2017).

In a study to evaluate the differential effects of the alkaloid upon cell viability, DNA damage, and nucleus integrity in mouse primary spleen cells and mouse lymphocytic leukemic cells, L1210 showed that sanguinarine produces dose-dependent increase in DNA damage and cytotoxicity in both cells (Maiti and Kumar, 2010). Matkar et al. (2008) had studied the effect of sanguinarine in human colon cancer cell line and showed its ability to cause DNA single- and double-stranded break as well as increased level of 8-oxodeoxyguanosine and reported DNA damage consistent with colon cancer cell death.

B. Plant extracts as DNA protectants

The demand for herbal medicines is greater than ever due to their safety and enhanced efficacy with least or no side-effects. Phytomedicines have got good acceptance by society (Rawat and Shankar, 2003; Kaur et al., 2019). Medicinal plants are significant source of natural and also many of the synthetic drugs. They have been utilized in the treatment and prevention of diseases and for the promotion of good health since antiquity. In modern pharmacology, several drug molecules are derived from plant sources (Sood et al., 2009). In a study, the DNA damage inhibitory activities of ethanolic, ethyl acetate, chloroform, and aqueous extracts of *C. papaya* have been performed. From the results of UV spectrophotometric analysis, it was observed that pretreatment with extracts followed by treatment with the oxidant or pretreatment with oxidant followed by treatment with extracts had significant DNA protective and repair activity. These findings were further confirmed with agarose gel electrophoresis. The order of DNA damage inhibition potential in both cases was found to be aqueous>ethanol>ethyl acetate>chloroform (Kadri et al., 2016).

As an effective alternative, natural product-based therapeutics are being increasingly explored as reliable alternative source for radioprotection. The availability of vast natural dietary and time-tested medicinal resources on this planet have been the motivation behind exploring the possibility of developing efficient, economically viable, and clinically acceptable radioprotectors for human applications. A good number of plants and herbs have been explored for their radioprotective activities. Some are categorized as 'Radio-protectors' because they can ameliorate the detrimental effects caused to the normal cells and reduce the side-effects of radiation therapy and others are called

'Radiosensitisers' since they augment radiation-induced cell death caused to the tumor, and thus curtail the dose of radiation treatment (Arora et al., 2005).

Ganoderma lucidum occurring in South India has been evaluated for its radioprotective properties of the aqueous extract using single-cell gel electrophoresis (comet assay), protection of radiation-induced plasmid pBR322 DNA strand breaks. The aqueous extract of *G. lucidum* at a concentration of 50µg/ml along with the DNA during irradiation, rendered protection to the plasmid DNA to an extent of 89.53%. These findings indicated the radio-protective efficacy of mushroom extract. (Pillai1 et al., 2006). A study was conducted on a fraction of wheat sprouts rich in glycoside molecule, a powerful antioxidant molecule against Fenton's reagent-induced oxidative damage of plasmid (pBR322) DNA. The results demonstrated the protective effects of the plant extracts used (Falcioni et al., 2002). *Hippophae rhamnoides* or seabuckthorn is used extensively in Indian and Tibetan traditional medicine for the treatment of circulatory disorders, ischemic heart disease, hepatic injury, and neoplasia. A study conducted by Shukla et al. (2006) has evaluated the radioprotective potential of REC-1001, a fraction isolated from the berries of *Hippophae rhamnoides*. The effect of REC-1001 extract was to modulate radiation-induced DNA damage. The results of this study indicated that the REC-1001 extract of *H. rhamnoides* could protect mitochondrial and genomic DNA from radiation-induced damage, with the maximum protective effect as observed at the highest REC-1001 dose evaluated i.e.250 lg/ml. REC-1001 exerted this effect in a dose-dependent manner. It scavenged radiation-induced hydroxyl radicals, chemically generated superoxide anions, stabilized DPPH radicals, and reduced Fe3þ to Fe2þ. The polyphenols/ flavonoids present in the extract might be responsible for the free radical scavenging and DNA protection as afforded by REC-1001 (Shukla et al., 2006). DNA protective ability of methanolic extract of the hill toon, *Cedrela serrata* leaves was evaluated against hydroxyl radical from hydrogen peroxide using agarose gel electrophoresis. The results indicated that the extracts have a potential DNA protective property from free radicals (Perveen et al., 2012).

C. Isolated phytocompounds as DNA protectants

Several plants are rich sources of phytochemicals like carotenoids, polyphenols, organo-sulphurs, saponins, alkaloids, which include anthocyanins, flavonoids, quercetin, stilbenes, tannins, berbarine lignins, *etc* (Jagetia, 2007). Among these, several flavonoids (quercetin, orientin, myricetin-flavonol, luteolin-flavone, and epigallocatechin gallate-flavanol, rutin, naringin, *etc*.) have been reported as potent antioxidants with radioprotective abilities (Benkovic et al., 2008;Srivastava et al., 2016; Lee et al., 2017).

Ten phytochemicals has been isolated and identified from seeds of the *Casimroa edulis* tree, which grows in Mexico and Central America. This tree produces edible fruits known as zapote blanco. The ethyl acetate extract from these seeds inhibited mutagenicity induced by 7, 12-dimethylbenz [1] anthracene with Salmonella typhimurium strain TM677. The extract completely inhibited DMBA-induced preneoplastic lesions *in vitro* in mouse mammary gland organ culture (Ito et al., 1998). Plants are a good source of bioactive compounds with negligible side-effects. These bioactive compounds play a significant role in terminating the generation of free radical chain reactions. *Bixa orellana* is a plant of interest from historic times due to different culinary and medicinal use. Hence, the DNA protective ability from oxidative stress of this plant was evaluated. The qualitative phytochemical analysis of different solvent extracts has revealed the presence of alkaloids, carbohydrates, tannins, phenolic compounds, carotenoids, flavonoids, diterpenes, and phytosterols. Flavonoids are the most important natural phenolic, and they possess a broad spectrum of chemical and biological activities including free radical scavenging properties. Flavonoids reduce free radicals by upregulating, quenching and have a variety of biological activities, such as antioxidative and cancer preventive. The presence of these phytocompounds such as tannins, flavonoids, carotenoids, and phenols in *Bixa orellana* seed extract may give credence to its local medicinal usage for different ailments (Kumar et al., 2014).

The ability of some natural products such as curcumin, resveratrol, indole-3-carbinol, and ellagic acid to modify the DNA-damaging ability of the alkylating carcinogen N-methyl-N'-nitro-N-nitrosoguanidine (MNNG) in cultured Chinese hamster lung fibroblast cells (CH V-79) is known. MNNG produced DNA single strand breaks in a dose- and time-dependent manner, as observed by increase in the tail moments of the comet, when the cells were subjected to alkaline single cell gel electrophoresis. When the cells were treated in the presence of each of the natural compounds, the DNA damage caused by MNNG was considerably reduced. This effect was found to be dose related. Preincubation of cells with each of these compounds individually afforded significant protection to DNA against damage caused by subsequent treatment with MNNG, indicating a true chemo-preventive role of these substances. The most remarkable aspect of the present study was that all four compounds helped in the recovery of DNA damage by accelerating DNA repair efficiency in the damaged cells. This was further substantiated by the observation on unscheduled DNA synthesis. Our results suggest that these agents are chemo-preventive by virtue of their ability to protect DNA as well as to induce DNA repair (Chakraborty et al., 2004).

Some specific neutraceuticals contained in daily diet have been investigated for their antioxidants and chemo-preventive efficiency. These include capsaicin (terpene found in green chilies and capsicum), curcumin (phenol found in rhizome of turmeric), ellagic acid (organic acid found in guava), fisetin (flavonoid found in strawberries), gallic acid (organic acid found in ginger), limolene (terpene found in citrus fruits like lemon and orange peel), lycopene (terpene found in tomato), quercetin (flavonoid found in citrus fruits, broccoli, onions, red grapes), resveratrol (polyphenol found in grapes) and rutin (flavonoid found in apricots, cherries, and citrus fruits). All these compounds were reported to have excellent antioxidants capacities and may have some chemo-preventive role against arsenic toxicity (Roy et al., 2008).

Alpha-tocopherol monoglucoside (TMG), a water-soluble derivative of α-tocopherol, has been examined for its ability to protect DNA against radiation-induced strand breaks. Gamma radiation, up to a dose of 6 Gy (dose rate, 0.7 Gy/ minute), induced a dose-dependent increase in single-strand breaks (SSBs) in plasmid pBR322 DNA. TMG inhibited the formation of radiation-induced DNA single strand breaks (SSBs) in a concentration-dependent manner (Rajagopalan et al., 2002).

CONCLUSION

A thorough survey of literature suggests that the phytochemicals exhibit potential to act as double-edged weapons i.e. to cause genotoxicity (DNA damage) as well as to protect DNA of different organisms. The underlying mechanisms of DNA damage by phytotoxicants, biotoxins, or any xenobiotics is the production of excessive ROS and generation of oxidative stress. In different cellular pathways, ROS in adequate amount are known to play a role as a signaling messenger but their higher concentrations lead to cellular damages in reversible or nonreversible manner. The oxidative stress thus produced has strong adverse impact on living cells which lead to emergence of different diseases including cancer. Several alkaloids, for example, display such DNA-damaging activities. In contrast, some of phytochemicals play a significant role in neutralizing ROS by acting as potentially protective antioxidants. The specific individual phytocompound or different molecules in combination may perform DNA protectant or antigenotoxic effects. Polyphenols, flavones, flavonoids, and carotenoids have been found to be associated with such functions. However, several *in vitro* and *in vivo* experiments are required to be carried out to ensure their safe clinical applications as efficient candidates to significantly reduce the level of DNA damage. In addition, extensive research is needed to properly understand the impact of phytochemicals on the DNA repair pathway.

REFERENCES

Ahmad, N., Gupta, S., Husain, M.M., Heiskanen, K.M., Mukhtar, H. 2000. Differential antiproliferative and apoptotic response of sanguinarine for cancer cells versus normal cells. *Clinical Cancer Research.* 6; 1524–1528.

Arora, R., Gupta, D., Chawla, R., Sagar, R., Sharma, A., Kumar, R., Prasad, J., Singh, S., Samanta, N., Sharma, R.K. 2005. Radioprotection by plant products: Present status and future prospects. *Phytotherapy Research* 19(1); 1–22.

Becci ,P.J., Schwartz, H., Barnes, H.H., Southard, G.L. 1987. Short-term toxicity studies of sanguinarine and of two alkaloid extracts of *Sanguinaria canadensis* L. *Journal of Toxicology and Environmental Health* 20(1–2); 199–208.

Benkovic, V., Knezevic, A.H., Dikic, D., Lisicic, D., Orsolic, N., Basic, I., Kosalec, I., Kopjar, N. 2008. Radioprotective effects of propolis and quercetin in gamma-irradiated mice evaluated by the alkaline comet assay. *Phytomedicine* 15(10); 851–858.

Chakraborty, S., Roy, M., Bhattacharya, R.K. 2004. Prevention and repair of DNA damage by selected phytochemicals as measured by single cell gel electrophoresis. *Journal of Environmental Pathology Toxicology and Oncology* 23(3); 215–226.

Croaker, A., King, G.J., Pyne, J.H., Anoopkumar-Dukie, S., Simanek, V., Liu, L. 2017. Carcinogenic potential of sanguinarine, a phytochemical used in 'therapeutic' black salve and mouthwash. *Mutation Research/Reviews in Mutation Research* 774; 46–56.

Falcioni, G., Fedeli, D., Tiano, L., Calzuola, I., Mancinelli, L., Marsili, V., Gianfranceschi, G. 2002. Antioxidant activity of wheat sprouts extract in vitro: Inhibition of DNA oxidative damage. *Journal of Food Science* 67(8); 2918–2922.

Habli, Z., Toumieh, G., Fatfat, M., Rahal, O.N., Gali-Muhtasib, H. 2017. Emerging cytotoxic alkaloids in the battle against cancer: Overview of molecular mechanisms. *Molecules* 22(2); 1–22.

Ito, A., Shamon, L.A., Yu, B., MataGreenwood, E., SangKook, L., van Breemen, R.B., Mehta, R.G., Farnsworth, N.R., Fong, H.H.S., Pezzuto, J.M., Kinghorn, A.D. 1998. Anti-mutagenic constituents of Casimiroa edulis with potential cancer chemopreventive activity. *Journal of Agriculture and Food Chemistry* 46(9); 3509–3516.

Jagetia, G.C. 2007. Radioprotective potential of plants and herbs against the effects of ionizing radiationm. *Journal of Clinical Biochemistry and Nutrition* 40(2); 74–81.

Jagetia, G.C., Rao. 2015. Isoquinoline alkaloid berberine exerts its antineoplastic activity by inducing molecular DNA damage in HeLa cells: A comet assay study. *Biology and Medicine (Aligarh)* 7; 1–7.

Kadri, S.U.T., Nikhitha, M., Shlini,. P. 2016. Cytoprotective and DNA Protective Activity of Carica Papaya Leaf Extracts. *International Journal of Pharmaceutical Science Invention* ISSN (Online): 2319 – 6718, ISSN (Print): 2319 – 670X www .ijpsi.org 5(4); 35–40.

Karp, J.M., Rodrigo, K.A., Pei, P., Pavlic, M.D., Andersen, J.D., McTigue, D.J., Fields, H.W., Mallery, S.R. 2005. Sanguinarine activates polycyclic aromatic hydrocarbon associated metabolic pathways in human oral keratinocytes and tissues. *Toxicology Letters* 158(1); 50–60.

Kaur, P., Purewal, S.S., Sandhu, K.S., Kaur, M. 2019. DNA damage protection: An excellent application of bioactive compounds. *Bioresources and Bioprocessing* 6(1), https://doi.org/10.1186/s40643-019-0237-9.

Kumar, Y., Babujestadi, D., Periyasamy, L. 2014. Preliminary phytochemical screening, DNA protection, antioxidant and antiproliferative effect of seed extracts of *Bixa orellana* L. *American Journal of Phytomedicine and Clinical Therapeutics*, *AJPCT* 2; 1024–1036.

Lee, M.T., Lin, W.C., Yu, B., Lee, T.T. 2017. Antioxidant capacity of phytochemicals and their potential effects on oxidative status in animals - A review. *Asian-Australasian Journal of Animal Sciences* 30(3); 299–308.

Lopus, M., and Panda, D. 2006. The benzophenanthridine alkaloid sanguinarine perturbs microtubule assembly dynamics through tubulin binding. A possible mechanism for its antiproliferative activity. *FEBS J.* 273, 2139–2150. doi: 10.1111/j.1742-4658.2006.05227.x

Maiti, M. and Kumar, G.S. 2010. 5Polymorphic Nucleic Acid Binding of Bioactive Isoquinoline Alkaloids and Their Role in Cancer. *Journal of Nucleic Acids* Volume 2010, Article ID 593408, 23 pages doi:10.4061/2010/593408

Matkar, S.S., Wrischnik, L.A., Hellmann-Blumberg, U. 2008. Sanguinarine causes DNA damage and p53-independent cell death in human colon cancer cell lines. *Chemico-Biological Interactions* 172(1); 63–71.

Motilal, M., Gopinatha, S.K. 2010. Polymorphic nucleic acid binding of bioactive isoquinoline alkaloids and their role in cancer. *Journal of Nucleic Acids* 2010; 1–23.

Niedzwiecki, A., Roomi, M.W., Kalinovsky, T., Matthias, Rath, M. 2016. Anticancer efficacy of polyphenols and their combinations. *Nutrients* 8(9); 552–568.

Perveen, F., Zaib, S., Irshad, S., Hassan, M., Perveen, F. et al. 2012. Antioxidant and DNA protection activities of the hill toon, *Cedrela serrata* (Royle) leaves extract and its fractions. *Journal of Natural Products* 5; 207–213.

Pillai, T.G., Salvi, V.P., Maurya, D.K., Nair, C.K.K., Janardhanan, K.K. 2006. Prevention of radiation-induced damages by aqueous extract of Ganoderma lucidum occurring in southern parts of India. *Current Science* 91; 341–344.

Rajagopalan, R., Wani, K., Huilgol, N., Kagiya, T.V., Nair, C.K.K. 2002. Inhibition of gamma-radiation induced DNA damage in plasmid pBR322 by TMG, a water-soluble derivative of vitamin E-Radiation Induced DNA Damage in Plasmid pBR322 by TMG. *Journal of Radiation Research* 43(2); 153–159.

Rawat, M.S., Shankar, R. 2003. Distribution status of medicinal plants conservation in Arunachal Pradesh with special reference to national medicinal plants board, *BMEBR* 24; 1–11.

Roy, M., Sinha, D., Mukherjee, S.C., Paul, S., Bhattacharya, R.K. 2008. Protective effect of dietary phytochemicals against arsenite induced genotoxicity in mammalian V79 cells. *Indian Journal of Experimental Biology* 46(10); 690–697.

Sarkar, S.N. 1948. Isolation from Argemone Oil of dihydrosanguinarine and sanguina rine: Toxicity of sanguinarine. *Nature* 162(4111); 265–266.

Shahi, N., Sahoo, M., Mallik, S.K., Sarma, D., Das, P. 2012. The microcystins-induced DNA damage in the liver and the heart of zebrafish, *Danio rerio*. *Toxicological and Environmental Chemistry* 94(2); 340–349.

Shukla, S.K., Chaudhary, P., Kumar, I.P., Samanta, N., Afrin, F., Gupta, M.L., Sharma, U.K., Sinha, A.K., Sharma, Y.K., Sharma, R.K. 2006. Protection from radiation-induced mitochondrial and genomic DNA damage by an extract of *Hippophae rhamnoides*. *Environmental and Molecular Mutagenesis* 47(9); 647–656.

Singh, N., Sharma, B. 2018a. Biotoxins mediated DNA damage and role of phytochemicals in DNA potection. *Biochemistry and Molecular Biology Journal* 4; 1–4.

Singh, N., Sharma, B. 2018b. Toxicological effects of berberine and sanguinarine. *Frontiers in Molecular Biosciences* 5; 1–7.

Sood, S., Bansal, S., Muthuraman, A., Gill, N.S., Bali, M. 2009. Therapeutic potential of *Citrus medica* L. peel extract in carrageenan induced inflammatory pain in rat. *Research Journal of Medicinal Plant* 3; 123–133.

Srivastava, S., Somasagara, R.R., Hegde, M., Nishana, M., Tadi, S.K., Srivastava, M., Choudhary, B., Raghavan, S.C. 2016. Quercetin, a natural flavonoid interacts with DNA, arrests cell cycle and causes tumor regression by activating mitochondrial pathway of apoptosis. *Scientific Reports* 6; 24049, https://doi.org/10.1038/srep24049.

Thota, S., Rodrigues, D.A., Barreiro, E.J. 2018. Recent advances in development of polyphenols as anticancer agents. *Mini-Reviews in Medicinal Chemistry* 18(15); 1–5.

Zegura, B., Gajski, G., Straser, A., Garaj-Vrhovac, V., Filipič, M. 2011. Microcystin-LR induced DNA damage in human peripheral blood lymphocytes. *Mutation Research* 726(2); 116–122.

Zhang, Y.J., Gan, R.Y., Li, S., Zhou, Y., Li, An-Na, Xu, D.P., Li, H.B. 2015. Antioxidant phytochemicals for the prevention and treatment of chronic diseases. *Molecules* 20(12); 21138–21156.

Role of phytochemicals in protection of DNA from damage by different environmental factors. A. Physical factors (Thermal and radiation). B. Chemical factors (Organic and inorganic). C. Biological factors (Plants and animals)

8

INTRODUCTION

Environment plays a crucial role in people's physical, mental, and social well-being. There are several abiotic (ambient temperature, intensity of sunlight, and pH of water and soil) and biotic (living organisms) environmental factors which keep influencing the biochemical and physiological status of any living cell. Some of these physical factors such as temperature or heat and radiation have abilities to modulate DNA structure and function through generation of ROS and RNS. In order to mitigate the adverse effects of these environmental factors and to safeguard the cellular health, various plant products have been used in the past. The abiotic environmental factors include (1) physical factors such as heat and radiations such as IR and UV / medical X-rays; and (2) the chemical agents such as pesticides, heavy metals, cigarette smoke, some food additives, chemotherapeutic drugs, solvents, air born pollutants, and industrial chemical etc. The environmental biotic factors comprise toxins derived from animal and plant sources (mainly their secondary metabolites or phytochemicals). The complex relationships between environmental factors and human health have been reported by several researchers (Seymour, 2016).

Phytochemicals are considered as a powerful group of compounds, belonging to secondary metabolites of plants. This group includes a diverse range of chemical entities such as flavonoids, polyphenols, organo-sulfur compounds, steroidal saponins, and vitamins. Plants have always been considered as a primary source of food and medicinal compounds. In a report, it has been mentioned that up to 200 species are considered as medicinal plants and approximately 25% of the total medicines have been derived from plants. Most part of edible plants are often described as nutraceuticals in literature. Nutraceuticals have achieved an important status due to their health-promoting efficiencies, including the prevention and treatment of several pathologic conditions. Vitamin C (ascorbic acid) was the first molecule to be discovered as an antioxidant. It is known to react rapidly with superoxide radicals O^{2-}, singlet oxygen and ozone (chemically), and H_2O_2 (enzymatically) through ascorbate peroxidase to neutralize the free radical species generated due to exposure to toxins. In plants, this acid is involved in the re-synthesis of carotenoids and vitamin E (tocopherol), which act as antioxidants and also provide protection from lipid peroxidation (Wilhelm and Helmut 2003; Forni et al., 2019). Fruits and vegetables are rich sources of dietary phytochemicals including phenolic acids and flavonoids (Liu, 2013).

These phenolic compounds have been considered as natural secondary plant metabolites, which mainly involve in the defense mechanism and provide protection to the organisms from abiotic or biotic stressors. Sivakumar et al. (2018) have indicated the use of traditional vegetables by the people from sub-Saharan Africa because of the presence of nutrient bioactive compounds which consist of health-promoting and protective agents. Carotenoids, present as red, orange, and yellow colors in flowers, leaves, vegetables, and fruits contain plenty of strong antioxidative properties (Wilhelm and Helmut, 2003). Their antioxidant activity is based on their scavenging peroxyl radicals.

The number of conjugated double bonds present in these molecules (α- and β-carotene) is related to the strength and the efficiency of quenching. Some polar carotenes such as zeaxanthin and cryptoxanthin belong to the group of highly active quenchers of $^1O_2$. The antioxidant properties of different carotenes isolated from sources have been reviewed by Stahl and Sies (2003). Synergistic effects in scavenging reactive nitrogen species (RNS) have been reported for β-carotene and vitamins C and E (Forni et al., 2019; Fiedor and Burda, 2014).

The most promising health-promoting molecules for further studies are phenolic compounds. These phytochemicals comprise a wide range of molecules (about 8000 different structures) playing important roles in the life of plants, where they are widely distributed in their different parts. These compounds can be divided into phenolic acids, lignans, lignins, stilbenes, tannins, and flavonoids. In plants, phenolics are involved in H_2O_2 detoxification, providing protection against UV radiation, also acting as enzyme modulators and feeding deterrents for herbivores. The broad spectrum of biological activities of phenolics, which contain the antioxidant (i.e., reducing agents, free radical scavenger, and quenchers of single oxygen formation) and antitumor properties, is widely acknowledged in several studies. The presence of at least one phenol ring is important for such activity with substituent by hydroxyl, methyl, or acetyl groups replacing the hydrogen. An increased antioxidant activity has been related to the enhanced number of free hydroxyls and conjugation of side chains to aromatic rings. Other potential antioxidants include the flavonoids containing several molecules of varying structures which could be categorized into six subclasses such as favonols, favones, favanones, favan-3-ols, isofavones, and anthocyanidins (Forni et al., 2019). These molecules are associated with strong antioxidant, antibacterial, anticancer, cardioprotective, immune booster, and anti-inflammatory properties. These molecules are found to protect the skin from highly dangerous UV radiation (Tungmunnithum et al., 2018).

PHYSICAL FACTORS (THERMAL AND RADIATIONS) INDUCED DNA ASSAULTS AND AMELIORATION BY PLANT PRODUCTS

Physical environmental factors which impact the structure of DNA are mainly radiations and heat. Several workers have reported DNA-damaging potential of both of these stressors. People are exposed to natural sources of ionizing radiations, such as in soil, water, and vegetation, as well as in human-made sources, such as X-rays and medical devices. The extent of damage to the tissues or organs depends on the duration of exposure and the dose of radiation given to an individual, or the dose of radiation absorbed. One unit of radiation is called gray (Gy). The severity of damage by any radiation absorbed depends not only on the radiation type but also on the sensitivity of varying tissues or organs. However, even the exposure organisms to ionizing radiation in low doses but for longer duration can increase the chances of development of cancer (WHO, 2016). Heat stress can also have a major influence on daily human activities. Heat stress affects mood, increases psychological distress, and mental health problems and also reduces key human psychological performance (Tawatsupa et al., 2012). In addition, heat stress has been reported to induce severe negative impact on DNA. Radiations are known to cause DNA damage via inducing production of free radicals and modulating the redox system of a living system (Lomax et al., 2013).

There are many reports available which suggest that scavenging of radiation-induced free radicals and elevation of cellular antioxidants could be mediated by the presence of variety of phytochemicals. Among all phytochemicals, polyphenols largely play key roles in chelating free radicals and neutralize them. The exposure of the living systems to ionizing radiations reduces the level of cellular antioxidants (both the enzymatic and non-enzymatic) and causes damage to genomic DNA. Several studies regarding protection against radiation-induced damage have been conferred by the up-regulation of DNA repair genes, which brings about an error-free repair of DNA damage. In a study, the ability of *Centella asiatica* (*C. asiatica*) extract has been evaluated for its DNA protection potential against the deleterious effects of ionizing radiation exposure. The results of their studies displayed that the treatment with this plant extract reduced radiation-induced damage to DNA significantly. Their findings suggest that radioprotection by *C. asiatica* extract could be mediated by mechanisms that act in a synergistic manner, especially involving antioxidant activity (Joy and Nair, 2009). The aqueous extract of a natural herb, *Terminalia chebula* (*T. chebula*), has been investigated for its antimutagenic activity by following the inhibition of gamma-radiation-induced strand breaks formation in plasmid pBR322 DNA. The results concluded that the aqueous extract of *T. chebula* protect cellular organelles from the radiation-induced damage. Researchers have proposed it to be considered as a probable radioprotector (Naik et al., 2004; Menon and Nair, 2013). In 2009, Rao et al. have indicated that the hydroalcoholic extract of *Cymbopogon citrates* exhibited the ability to prevent DNA damage by radiation in V79 cells as observed by the micronucleus assay. These authors also demonstrated the free radical-quenching activity when tested *in vitro*. Jagetia et al. (2012) have demonstrated in mice brain that the extract of jamun *(Syzygium cumini)* has the ability to significantly reduce lipid peroxidation in a concentration-dependent manner in cell culture systems. The results of their study further demonstrated in a cell-free system that the extract of jamun could inhibit the formation of reactive oxygen species such as O(2)-, OH, DPPH, and ABTS(+) free radical; the effect being concentration dependent. They observed that this plant extract could protect DNA from damage by y-radiation.

Commercially available Ayurvedic formulations, Brahma Rasayana (BRM) and Chyavanaprash (CHM), have been analyzed for their ability to restore cellular antioxidant status and enhance repair of radiation-induced DNA damages in mice (orally administered) following whole body exposure to gamma radiation. This work suggests the possibility of using BRM or CHM as a therapeutic radioprotector during unplanned, accidental ionizing radiation exposure scenario (Menon and Nair, 2013). Some phenolic compounds were isolated from the whole plant of *Pilea microphylla* using conventional opensilica gel column chromatography and preparative HPLC. These phenolic compounds were quercetin-3-O-rutinoside, 3-O-caffeoylquinic acid, luteolin-7-O-glucoside, apigenin-7-O-rutinoside, apigenin-7-O-b-D-glucopyranoside, and quercetin. These tested compounds were found to significantly prevent the Fenton reagent-induced calf thymus DNA damage (Bansala et al., 2011). The radioprotective effect of methanolic extract (10, 50, 100, and 200 µg/mL) of *Achillea millefolium* L (ACM) for 2h was examined against genotoxicity induced by ionizing radiation (IR) in human lymphocytes. At each dose point, whole blood was exposed *in vitro* to 2.5 Gy of X-ray. The maximum protection and decrease in frequency of micronuclei formation were observed at 200 µg/mL of ACM extract which completely protected genotoxicity induced by IR in human lymphocytes. These observations suggested that the methanolic extract of ACM might play an important role in the protection of normal tissues against genetic damage induced by IR (Shahani et al., 2015). Aqueous extract of *Tinospora cordifolia*, a plant from the Menispermaceae family, has been found to inhibit radiation mediated 2-deoxyribose degradation in a dose-dependent manner. The radio sensitization activity of this extract may be due to elevated levels of DNA damage of tumor cells. Thus, the use of this extract may offer an alternative treatment strategy for cancer in combination with gamma radiation (γ-GR). This plant is known to synthesize some important bioactive secondary metabolites such as berberine, magnoflorine, palmatine, tembetarine, choline, β-sitosterol, hydroxyl ecdysone, ecdysterone, isocolumbin, magnoflorine, palmatine, tetra-hydro-palmatine, and tinosporin.

On the other hand, the effect of heat stress or heat shock or heperthermia has also been studied extensively in many organisms (Vidhya et al., 2018) but the exact mechanism of cellular response to it involving cellular compartments and metabolic pathways is still not known (Kantidze et al., 2016). The involvement of heat stress in inducing oxidative stress has received much attention. The generation of oxidative stress due to heat exposure can manifest in all parts of the body; mainly causing mitochondrial dysfunction, generation of excessive free radicals, and finally oxidative stress (Akbariyan et al., 2016). Earlier it has been demonstrated that exposure to heat stress may increase sensitivity to those factors responsible for causing double-stranded DNA breaks (Iliakis et al., 2008). This event is known as 'heat radiosensitization'. This effect manifests because of inhibition of basically two different DNA repair systems: (1) base excision repair system (Batuello et al., 2009; Fantini et al., 2013) and (2) the nucleotide excision repair system (Hettinga et al.,1997; Muenyi et al., 2011); the former being extensively studied. It has been observed that the activities of DNA polymerase β (Dikomey et al., 1987), some DNA glycosylases (Fantini et al., 2013) and mismatch repair system (Nandin etal, 2012) are significantly inhibited by exposure to the heat stress. Both of the DNA repair systems, i.e. non-homologous DNA end joining (NHEJ) and the homologous recombination (HR), are inhibited by heat stress (Oei et al., 2015; Kantidze et al., 2016).

Some phytochemicals, such as flavonoids and other related compounds, have been shown to be useful in inhibiting chronic heat-stressed DNA damage in poultry, but were less or not effective in non-heat-stressed counterparts. This finding supports the notion that phytochemicals have potential antioxidant properties under challenging conditions. Though significant amount of work has been conducted to clarify our understanding about the relationship between oxidative stress and heat, the exact pathway(s) of actions of phytochemicals to encounter the condition of oxidative stress is still not clear (Akbarian et al., 2016). These authors have explained that heat stress is a non-specific physiological response of the body when exposed to high ambient temperatures, which can break the balance of redox system of the body thereby developing the oxidative stress which adversely influences growth performance and poultry health. Recently, polyphenols have gained huge attention due to their antioxidant potential and hence, they serve for heat stress as efficient attenuators. These studies have demonstrated the free radical generation potential of heat stress, followed by DNA dama So, the production of ROS and RNS due to heat stress could significant contribute to induce alterations in the structure of DNA and also cause it damage. They have also demonstrated that introduction of phytochemicals can recover DNA from damage or any other adverse effects of heat stress (Hu et al., 2019).

CHEMICALS (ORGANIC AND INORGANIC) MEDIATED DNA ASSAULTS AND AMELIORATION BY PLANT EXTRACTS

Several chemical agents have been reported to act as strong mutagens with enough potential to cause DNA damage significantly (Aurbasch, 1976; Barnes et al., 2018). Heterocyclic aromatic amines (HAAs) and polycyclic aromatic hydrocarbons (PAHs) are the main groups of chemicals recognized for their abilities to cause DNA damage (Barnes et al., 2018). However, not all of the anthropogenic (man-made) chemicals are able to cause DNA damage within living systems. Many of them remain present in inactivated or less active form in native state and they need to undergo metabolic transformations to get converted into highly active form, a process called bioactivation.

In addition, there are many renewable and non-renewable sources of energy such as coal, gases, and tobacco which undergo combustion and produce various secondary pollutants such as PAAs; some of them cause DNA damage (Noah et al., 2019; Smit et al., 2019; Aucella et al., 2019; Guibert et al., 2019). Smoking, which contains several carcinogens, is solely a man-made DNA-damaging exogenous agent which is highly toxic in nature. Smoking plays an important role in formation of tumor as well as initiation of lung cancers (Rojewski et al., 2018; Kaufman et al., 2018; Christensen et al., 2018; Donner et al., 2018).

The natural resources with potent antioxidants and bioactive constituents should be analyzed in an array of different model systems. The efficacy of bioactive constituents depends on soil profile, moisture content, stress conditions, and nutrient absorption rate by the natural resources under a given set of conditions. However, suitable analytical protocols need to be adopted to monitor the kinetics of antioxidants during OS conditions. Extracts prepared from specific natural resources are currently being used to evaluate their

A protecting efficacy from oxidative damage using Fenton's reagent. The mechanism involved in DNA damage underlies the breakdown of one of the phosphodiester chains in response to Fenton's reagent. In response to hydroxyl radicals, the supercoiled form of DNA changes into a relaxed form. Fenton's reagent-mediated reaction complex leads to the formation of hydroxyl radical in the presence of hydrogen peroxide (H_2O_2) and Fe^{3+}. Breakdown of DNA strand may cause chronic diseases and age-related medical issues (Gao et al., 2014; Guleria et al., 2017).

The association of diet, stress conditions, and DNA repair is of utmost importance. The significant DNA damage protection ability of extracts of Amazon moss (Fernandes et al., 2018), Ashtvarga (Giri et al., 2017), Bael flower (Chandrasekara et al., 2018), Barley (Salar et al., 2017a), Carissa carandas leaves (Verma et al., 2015), Eulophia nuda Lindl (Kumar et al., 2013), *Garcinia gracilis* leaves (Supasuteekul et al., 2016), Grape seeds (Aybastier et al., 2018), Honey (Habib et al., 2014), Mung beans (Xio et al., 2015), pearl millet (Salar et al., 2017b), Sphallerocarpus gracilis seeds (Gao et al., 2014), Sugarcane, (Abbas et al., 2014); Teucrium polium, and Stachys iberica (Tepe et al., 2011) have been reported. Their immense antioxidant potential has been attributed toward their properties of inhibition of DNA damage.

Natural resources such as fruit, vegetables, cereal grains, and medicinal plants with antioxidant properties have played a significant role in providing better health to organisms. They have been considered as rich sources of bioactive components including vitamins, antioxidants, anthocyanins, sterols, and minerals. The bitter component (Naringenin) present in citrus fruits has shown promising results as anticancerous agents. Their chemo-sensitizing properties have also been reported. Naringenin inhibited azoxymethane-induced colon carcinogenesis and protected plasmid DNA from UVB-induced DNA damage. Antioxidant-rich natural resources are grown all over the world for their health benefiting secondary metabolites and functional products. The multifunctional bioactive components present in natural resources are solvent specific and sensitive for extraction parameters (extraction phase, temperature and time).

Bioactive components such as phenolics, flavonoids, tannins and other important phytochemicals (saponins, catechol tannin, anthocyanin, steroids and sugars) have been shown to possess plenty of antioxidant properties. Quercetin is an important bioactive constituent with antioxidant properties which has shown many health benefits including cancer prevention and treatment. Available scientific reports indicate that quercetin has abilities to interact with human telomerase sequences and to stabilize the G-quadruplex structure. The binding of flavonoids with DNA duplex offers protection of DNA from the oxidative damage. In a cell culture experiment, it was observed that Quercetin can negatively influence the rate of formation ‹ ROS, thus reducing the number of lesions in PC12 neuronal cells. Quercetir arrests the cell cycle at some specific sites which causes apoptosis via mitochondrial pathway.

AMELIORATION OF DNA DAMAGE CAUSED BY BIOLOGICAL AGENTS

In addition to man-made chemicals, certain microorganisms and plant products are also responsible for causing DNA damage in different organisms. Some fungal strains produce certain metabolites or biochemicals under harsher conditions or in response to specific media components which contribute to generate certain cancers in humans. One of such fungal metabolites is aflatoxins, especially AFT-B1 produced by Aspergillus spp., which acts as hepatocarcinogen and results in abnormality and cirrhosis in liver. A reactive epoxide may be formed in response to AFT-B1 which reacts at specific position (N7) of guanine and causes metabolic dysfunction of liver. In addition to fungal species, certain plants synthesize and secrete secondary compounds in response to stress conditions which are reported to act like carcinogens. Among many carcinogens, some plant-based compounds such as limonen, aristolochia acid, reserpin, and arecolin may cause cancer. Safrol is another important compound found in pepper, celery, and *Sassafras albidum* which needs to be studied in detail for its carcinogenic potential.

Some of the bio-antioxidants such as riboflavin and α-tocopherol have been demonstrated to protect Epidemic Dropsy patients. A mixture of riboflavin and α-tocopherol has been evaluated against argemone oil (AO) and sanguinarine-induced genotoxicity using alkaline comet assay. Some workers carried out a study using single dose of a mixture of α-tocopherol (150 mg/kg) and riboflavin (50 mg/kg) to the experimental mice, 24h prior to or soon after the AO (2.0 ml/kg) exposure. The results from comet assay indicated significant reduction in the tail movement (70–72%), tail length (37–44%), and tail DNA (49–53%) in bone marrow cells. When they introduced antioxidants in single or multiple doses after 24h of treatment with AO, the comet assay results showed substantial (P 0.05) reduction in above parameters in the bone marrow cells. They used antioxidants in single dose and introduced it into mice either 24h before or soon after exposure to sanguinarine (21.6 mg/kg). The results showed that this treatment resulted in significant reduction in the movement of tail (56–62%), length of tail (69%) and the DNA tail (34–42%) in bone marrow cells from mice. These workers found that the application of antioxidants in the single dose or multiple doses after 24h of treatment with sanguinarine caused reduction in the movement of tail (50–71%), length of tail (54–63%) and DNA of tail (29–43%) in the cells from bone marrow. They obtained similar protective response after using a combination of antioxidants with mice blood cells exposed to AO or sanguinarine. On the basis of these results, they suggested that a mixture of α-tocopherol and riboflavin may offer significant protection against AO and sanguinarine-mediated DNA damage (Ansari et al., 2006).

CONCLUSION

Environmental factors responsible for causing DNA damage include both abiotic and biotic factors. Abiotic agents include different kinds of radiations (gamma rays and alpha particles), heat stress, and several anthropogenic chemicals (both inorganic such as As, Ni, Cd,Ni, asbestos (silicate minerals) etc. and organic dioxins, alkaloids, polycyclic aromatic hydrocarbons or PAHs, heterocyclic aromatic amines or HAAs, etc.). Biotic factors causing DNA damage include those chemicals synthesized and released into the environment by different organisms (both plants and animals) such as Aflatoxin B synthesized by the fungus *Aspergillus flavus* growing on stored nuts, peanut butter and grains; microbial toxins produced by *Helicobacter pylori*, helminthic products from *Opisthorchis viverrini* and *Clonorchis sinensis* and viruses (hepatitis B, and human papilloma virus, Peyton Rous, Rous sarcoma virus etc.). Each of these factors induces DNA damage through specific mechanisms and generates different types of cancers in humans via their direct or indirect interventions in the programmed cell death. In addition to these known carcinogens, there are some co-carcinogens which on their own do not cause cancer but modulate the activities of other chemicals so as to generate cancer. Since DNA molecules are nucleophiles, they are always susceptible to be attacked by electrophilic chemical agents. Several molecules synthesized as secondary metabolites such as polyphenols, flavonoids, flavones, carotenes, and terpenes etc., have been found to protect cellular DNA from xenobiotics-mediated damage primarily via quenching free radicals and thus reducing lipid peroxidation and oxidative stress. Extensive research, however, is needed to explore phytochemicals in order to reap their optimum benefits.

REFERENCES

Abbas, S.R., Sabir, S.M., Ahmad, S.D., Boligon, A.A., Athayde, M.L. 2014. Phenolic profile, antioxidant potential and DNA damage protecting activity of sugarcane (Saccharum officinarum). *Food Chemistry* 147; 10–16. https://doi.org/10.1016/j .foodchem.2013.09.113

Akbarian, A., Michiels, J., Degroote, J., Majdeddin, M., Golian, A., Smet, S.D. 2016. Association between heat stress and oxidative stress in poultry; mitochondrial dysfunction and dietary interventions with phytochemicals. *Journal of Animal Science and Biotechnology* 7(37); 1–14.

Ansari, K.M., Dhawan, A., Khanna, S.K., Das, M. 2006. Protective effect of bioantioxidants on argemone oil/sanguinarine alkaloid induced genotoxicity in mice. *Cancer Letters* 244(1); 109–118.

Aybastier, O., Dawbaa, S., Demir, C. 2018. Investigation of antioxidant ability of grape seeds extract to prevent oxidatively induced DNA damage by gas chromatography-tandem mass spectrometry. *Journal of Chromatography B* 1072; 328–335. https://doi.org/10.1016/j.jchromb.2017.11.044

Aucella, F., Prencipe, M., Gatta, G., Aucella, F., Gesualdo, L. 2019. Environment, smoking, obesity, and the kidney. In: *Critical Care Nephrology* (Third Edition). pp. 1320–1324.e1 https://doi.org/10.1016/B978-0-323-44942-7.00221-1

Auerbach C. 1976. Chemical mutagens: alkylating agents. II: Chemistry. Molecular analysis of mutants. Influence of cellular processes. Kinetics. In: *Mutation Research*. Springer, Boston.

Bansala, P., Paula, P., Nayaka, P.G., Pannakale, S.T., Zoud, J., Laatschd, H., Priyadarsinie, K.I., Unnikrishnan, M.K. 2011. Phenolic compounds isolated from Pilea microphylla prevent radiation-induced cellular DNA damage. *Acta Pharmaceutica Sinica B* 1(4); 226–235.

Barnes, J.L., Zubair, M., John, K., Poirier, M.C., Martin, F.L. 2018. Carcinogens and DNA damage. *Biochemical Society Transactions* 46 (5); 1213–1224.

Batuello, C.N., Kelley, M.R., Dynlacht, J.R.. 2009. Role of Ape1 and base excision repair in the radioresponse and heat-radiosensitization of HeLa Cells. *Anticancer Research* 29; 1319–1325.

Chandrasekara, A., Daugelaite, J., Shahidi, F. 2018. DNA scission and LDL cholesterol oxidation inhibition and antioxidant activities of Bael (Aegle marmelos) flower extracts. *Journal of Traditional and Complementary Medicine* 8;428–435. https ://doi.org/10.1016/j.jtcme.2017.08.010

Christensen, N.L., Lokke, A., Dalton, S.O., Christensen, J., Rasmussen, T.R. 2018. Smoking, alcohol, and nutritional status in relation to one-year mortality in Danish stage I lung cancer patients. *Lung Cancer* 124:40–44. https://doi.org/10 .1016/j.lungcan.2018.07.025

Dikomey, E., Becker, W., Wielckens, K. 1987. Reduction of DNA-polymerase beta activity of CHO cells by single and combined heat treatments. *International Journal of Radiation Biology and Related Studies in Physics, Chemistry, and Medicine* 52; 775–785.

ner, J., Anderson, H., Davison, S., Hughes, A.M., Bouirmane, J., Lindqvist, J., Lytle, K.M., Ganesan, B., Ottka, C., Ruotanen, P., Kaukonen, M., Forman, O.P., Fretwell, N., Cole, C.A., Lohi, H. 2018. Frequency and distribution of 152 genetic disease variants in over 100,000 mixed breed and purebred dogs. *PLoS Genet.* 14(4); e1007361.

Fantini, D., Moritz, E., Auvre, F., Amouroux, R., Campalans, A., Epe, B., Bravard, A., Radicella, J.P. 2013. *DNA Repair (Amst.).* 12(3); 227–237.

Fernandes, A. S., Mazzei, J.L., Evangelista, H., Marques, M.R.C., Ferraz, E.R.A., Felzenszwalb, I. 2018. Protection against UV-induced oxidative stress and DNA damage by Amazon moss extracts. *Journal of Photochemistry and Photobiology B* 183; 331–341. https://doi.org/10.1016/j.jphotobiol.2018.04.038

Fiedor, J., Burda, K. 2014. Potential role of carotenoids as antioxidants in human health and disease. *Nutrients* 6(2); 466–488.

Forni, C., Facchiano, F., Bartoli, M., Pieretti, S., Facchiano, A., D'Arcangelo, D., Norelli, S., Valle, G., Nisini, R., Beninati, S., Tabolacci, C., Jadeja, R.N. 2019. Beneficial role of phytochemicals on oxidative stress and age-related diseases. *BioMed Research International* 2019; 1–16.

Gao, C.Y., Tian, C., Zhou, R., Zhang, R., Lu, Y. 2014. Phenolic composition, DNA damage protective activity and hepatoprotective effect of free phenolic extract from Sphallerocarpus gracilis seeds. *International Immunopharmacology* 20;238–247. https://doi.org/10.1016/j.intimp.2014.03.002

Giri, L., Belwal, T., Bahukhandi, A., Suyal, R., Bhatt, I.D., Rawal, R.S., Nandi, S.K. 2017. Oxidative DNA damage protective activity and antioxidant potential of Ashtvarga species growing in the Indian Himalayan Region. *Industrial Crops and Products* 102; 173–179. https://doi.org/10.1016/j.indcrop.2017.03.023

Guilbert, A., Cremer, K.D., Heene, B., Demoury, C., Aerts, R., Declerck, P., Brasseur, O., Nieuwenhuyse, A.V. 2019. Personal exposure to traffic-related air pollutants and relationships with respiratory symptoms and oxidative stress: a pilot cross-sectional study among urban green space workers. *Science of the Total Environment* 649; 620–628. https://doi.org/10.1016/j.scitotenv.2018.08.338

Guleria, S., Singh, G., Gupta, S., Vyas, D. 2017. Antioxidant and oxidative DNA damage protective properties of leaf, bark and fruit extracts of *Terminalia chebula*. *Indian Journal of Biochemistry and Biophysics (IJBB)* 54; 127–134. http://nopr.niscair.res.in/handle/123456789/43106

Habib, H.M., Al-Meqbali, F.T., Kamal, H., Souka, U.D., Ibrahim, W.H. 2014. Bioactive components, antioxidant and DNA damage inhibitory activities of honeys from arid regions. *Food Chemistry* 153; 28–34. https://doi.org/10.1016/j.foodchem.2013.12.044

Hettinga, J.V., Konings, A.W., Kampinga, H.H. 1997. Reduction of cellular cisplatin resistance by hyperthermia--a review. *International Journal of Hyperthermia* 13(5); 439–457.

Hu, R., He, Y., Arowolo, M.A., Wu, S., He, J. 2019. Polyphenols as potential attenuators of heat stress in poultry production. *Antioxidants* 8; 67.

Iliakis, G., Wu, W., Wang, M. 2008. DNA double strand break repair inhibition as a cause of heat radiosensitization: Re-evaluation considering backup pathways of NHEJ. *International Journal of Hyperthermia* . 24(1); 17–29.

Jagetia, G.C., Shetty, P.C., Vidyasagar, M.S. 2012. Inhibition of radiation-ind DNA damage by jamun, Syzygium cumini, in the cultured splenocytes of m exposed to different doses of γ-radiation. *Integrative Cancer Therapies* 11(. 141–153.

Joy, J., Nair, C.K. 2009. Protection of DNA and membranes from gamma-radiation induced damages by centella asiatica. *Journal of Pharmacy and Pharmacology* 61(7); 941–947.

Kantidze, O.L., Velichko, A.K., Luzhin, A.V., Razin, S.V. 2016. Heat stress-induced DNA damage.*Acta Naturae* 8(2); 75–78.

Kaufman, A.R., Dwyer, L.A., Land, S.R., Klein, W.M.P., Park, E.R. 2018. Smoking-related health beliefs and smoking behavior in the National Lung Screening Trial. *Addictive Behaviors* 84; 27–32.

Kumar, A., Lemos, M., Sharma, M., Shriram, M. 2013. Antioxidant and DNA damage protecting activities of Eulophia nuda Lindl. *Free Radicals and Antioxidants* 3;55–60. https://doi.org/10.1016/j.fra.2013.07.001

Liu, R.H. 2013. Health-promoting components of fruits and vegetables in the diet. *Advances in Nutrition* 4(3); 384S–392S. doi: 10.3945/an.112.003517

Lomax, M.E., Folkes, L.K., O'Neill, P. 2013. Biological consequences of radiation-induced DNA damage: Relevance to radiotherapy. *Clinical Oncology* 25(10); 578–585.

Menon, A., Nair, C.K.K. 2013. Ayurvedic Formulations as therapeutic radioprotectors: Preclinical studies on brahma rasayana and chyavanaprash. *Current Science* 104(7); 959–966.

Muenyi, C.S., States, V.A., Masters, J.H., Fan, T.W., Helm, C.W., States, J.C. 2011. Sodium arsenite and hyperthermia modulate cisplatin-DNA damage responses and enhance platinum accumulation in murine metastatic ovarian cancer xeno-graft after hyperthermic intraperitoneal chemotherapy (HIPEC). *Journal of Ovarian Research* 4; 1–11.

Nadin, S.B., Cuello-Carrion, F.D., Sottile, M.L., Ciocca, D.R., Vargas-Roig, L.M. 2012. Effects of hyperthermia on Hsp27 (HSPB1), Hsp72 (HSPA1A) and DNA repair proteins hMLH1 and hMSH2 in human colorectal cancer hMLH1-deficient and hMLH1-proficient cell lines. *International Journal of Hyperthermia* 28(3); 191–201.

Naik, G.H., Priyadarsini, K.I., Naik, D.B., Gangabhagirathi, R., Mohan, H. 2004. Studies. Oncology and therapy studies on the aqueous extract of Terminalia chebula as a potent antioxidant and a probable radioprotector. *Phytomedicine* . 11(6); 530–538.

Noah, S., Rozich, M.D., Alessandra, L.M.D., Casey, S., Butler, M.D., Morgan, M., Bonds, M.D., Laura, E., Fischer, M.D., Russell, G., Postier, M D., Katherine, T., Morris, M.D. 2019. Tobacco smoking associated with increased anastomotic disruption following pancreaticoduodenectomy. *The Journal of Surgical Research* 233; 199–206. https://doi.org/10.1016/j.jss.2018.07.047

Oei, A.L., Vriend, L.E., Crezee, J., Franken, N.A., Krawczyk, P.M.. 2015. Effects of hyperthermia on DNA repair pathways: one treatment to inhibit them all. *Radiation Oncology* 10; 165.

Rojewski, A.M., Tanner, M.D., Dai, L., Ravenel, M.D., Mulugeta, G., Silvestri, G.A., Toll, B.A. 2018. Tobacco dependence predicts higher lung cancer and mortality rates and lower rates of smoking cessation in the National Lung Screening Trial. *Chest* 154; 110–118. https://doi.org/10.1016/j.chest.2018.04.016

, B.S., Shanbhoge, R., Rao, B.N., Adiga, S.K., Upadhya, D., Aithal, B.K., Kumar, M.R. 2009. Preventive efficacy of hydroalcoholic extract of *Cymbopogon citratus* against radiation-induced DNA damage on V79 cells and free radical scavenging ability against radicals generated in vitro. *Human and Experimental Toxicology* 28(4); 195–202.

Seymour, V. 2016. The human–nature relationship and its impact on health: A critical review. Front Public Health 4: 260.

Shahani, S., Rostamnezhad, M., Ghaffari-rad, V., Ghasemi, A., Pourfallah, T.A., Hosseinimehr, S.J. 2015. Radioprotective effect of *Achillea millefolium* L against genotoxicity induced by ionizing radiation in human normal lymphocytes. Dose Response 13(1); 1–5.

Sivakumar, D., Chen, L., Sultanbawa, Y. 2018. A comprehensive review on beneficial dietary phytochemicals in common traditional southern African leafy vegetables. *Food Science and Nutrition* 6(4); 714–727.

Smit, T., Peraza, N., Garey, L., Langdon, K.J., Ditre, J.W., Rogers, A.H., Manning, K., Zvolensky, M.J. 2019. Pain-related anxiety and smoking processes: the explanatory role of dysphoria. *Addictive Behaviors* 88; 15–22. https://doi.org/10.1016/j.addbeh.2018.08.008

Stahl, W., Sies H. 2003. Antioxidant activity of carotenoids. *Molecular Aspects of Medicine* 24(6); 345–351.

Supasuteekul, C., Nonthitipong, W., Tadtong, S., Likhitwitayawuid, K., Tengamnuay, P., Sritularak, B. 2016. Antioxidant, DNA damage protective, neuroprotective, and -glucosidase inhibitory activities of a flavonoid glycoside from leaves of *Garcinia gracilis*. *Brazilian Journal of Pharmacognosy* 26; 312–320.

Tawatsupa, B., Yiengprugsawan, V., Kjellstrom, T., Seubsman, S., Sleigh, A. Heat stress, health and well-being: Findings from a large national cohort of Thai adults. doi:10.1136/bmjopen-2012-001396.

Tepe, B., Degerli, S., Arslan, S., Malatyali, E., Sarikurkcu, C. 2011. Determination of chemical profile, antioxidant, DNA damage protection and antiamoebic activities of Teucrium poliumand Stachys iberica. *Fitoterapia* 82; 237–246. https://doi.org/10.1016/j.fitote.2010.10.006

Tungmunnithum, D., Thongboonyou, A., Pholboon, A., Yangsabai, A. 2018. Flavonoids and other phenolic compounds from medicinal plants for pharmaceutical and medical aspects: An overview. *Medicines (Basel)* 5(3); 93.

Verma, K., Shrivastava, D., Kumar, G. 2015. Antioxidant activity and DNA damage inhibition in vitro by a methanolic extract of Carissa carandas (Apocynaceae) leaves. *The Journal of Taibah University for Sciences* 9; 34–40. https://doi.org/10.1016/j.jtusci.2014.07.001

Vidhya, V., Krishnamoorthy, M., Venkatesan, V., Jaganathan, V., Paul, S.F.D. 2018. Occupational HEAT STRESS, DNA damage and heat shock protein - a review. *Medical Research Archives* 6; 1–19.

WHO, 2016. Ionizing radiation, health effects and protective measures. Website: https://www.who.int/news-room/fact-sheets/detail/ionizing-radiation-health-effects-and-protective-measures.

Wilhelm, S.W., Helmut, S.H. 2003. Antioxidant activity of carotenoids. *Molecular Aspects of Medicine* 24(6); 345–351.

Xiao, Y., Zhang, Q., Miao, J., Rui, X., Li, T., Dong, M. 2015. Antioxidant a and DNA damage protection of mung beans processed by solid state ferm tion with *Cordyceps militaris* SN-18. *Innovative Food Science and Emerg Technologies* 31; 216–225. https://doi.org/10.1016/j.ifset.2015.06.006

DNA protective and genotoxic potential of environmental factors

Experimental models for the study

INTRODUCTION

Genetic factors are important in terms of influencing individual susceptibility to carcinogens. However, external factors represent the greatest opportunity for primary prevention. By 'external factors' we mean those related to the environment a broad scope, including all non-genetic factors such as diet, lifestyle, and infectious agents. In a more specific approach, environmental factors include natural or man-made agents encountered by humans in their daily life, upon which they have no or limited personal control (Hernandez and Blazer, 2006). Environmental chemicals including contaminant anthropogenic agents cause pollution of air, water, and soil. Occupational exposure of these chemicals to humans serves as potential threat to health and induces the development of disease. Also, the diet taken by a person plays an important role in carcinogenesis. Some workers have established the association between nutrition and some types of cancer. Biomarkers of genotoxicity have been widely exploited for detecting the early initiation of cancer in an individual. These biomarkers act as important tools for studying molecular epidemiology of cancer and also for its biomonitoring. Molecular epidemiology establishes association between the exposure of an individual to the toxic chemicals or the factors responsible for genetic susceptibility and the onset of the disease (Ladeira and Smajdova, 2017).

Some of the chemicals, e.g. pesticides and heavy metals, may be geno-toxic to the sentinel species and/or to non-target species, causing deleterious effects in somatic or germ cells. The test systems which help in the prediction of hazards and risk assessment are important to assess the genotoxic potential of chemicals in terms of DNA damage in flora and fauna before their release into the environment or commercial applications. The comet assay has been widely accepted as a simple, sensitive, and rapid tool for assessing DNA damage and repair in any eukaryotic and/or prokaryotic cells. The comet assay has tremendous application in diverse fields ranging from genetic toxicology to epidemiology (Hartwig et al., 2020).

This chapter deals with the comprehensive presentation of the use of different assays and varying experimental models from bacteria to man, and also employing diverse cell types to assess the DNA-damaging potential of chemicals and/ or environmental factors. Sentinel species are the first to be affected by adverse changes in their environment. Determination of DNA damage would provide information about the genotoxic potential of varied carcinogens at an early stage in any indicator organisms. This would permit for implementing immediate intervention strategies for prevention or reduction of deleterious health effects in the sentinel species as well as in humans.

ENVIRONMENTAL EXPOSURE TO GENOTOXICANTS

Organisms, due to their changing lifestyles and innovations, get continuously exposed to many mutagenic and carcinogenic agents released into the environment. Chemical agents may also include drugs, food additives, pesticides, and nanomaterials. Anthropogenic pollution has become a very common phenomenon to modern environment (Chaudhari and Saxena, 2016). The global and rapid increase in technogenic stress in the biosphere raises questions about possible consequences for biota, including man, acknowledging that all forms of life are interconnected and that human health is strongly linked to the ecosystem's health. Environmental chemicals and contaminants are ubiquitous, occurring in water, air, food, and soil. While some chemicals are short-lived in the environment and may elicit no harmful effects in humans, other chemicals bioaccumulate or persist for a long time in the environment or the human body due to frequent exposure, potentially leading to adverse health effects (Vassilev and Menendez, 2005; Rhind, 2009).

A more integrated approach is needed to deal with the fact that a biological effects induced by exposure to complex pollutant mixtures are easily interpreted from a set of chemical analyses. The toxic effect of differ interacting pollutants can be either additive, synergistic, or antagonistic. Th results obtained from several molecular epidemiology studies conducted on the environmentally or occupationally exposed populations to high concentrations of complex mixtures of several urban air pollutants have illustrated severe genotoxic effects in the form of enhanced incidence of DNA damage. Atmospheric pollutants such as carbon monoxide, ozone, nitrogen oxides, sulfur dioxide, polycyclic aromatic hydrocarbons, and particulate matter are the examples of chemical agents that may cause DNA damage and pose a serious threat to the health and well-being of humans and other organisms. According to their physicochemical properties, for instance, polycyclic aromatic hydrocarbons (PAHs) are released into the environment from both natural and anthropogenic sources, and are highly mobile in the environment, allowing them to distribute across air, soil, and water, becoming effectively ubiquitous. It is also of great importance to assess the risk of future health effects from accidental or occupational radiation exposure to humans in order to be able to take appropriate measures to protect exposed individuals. A multidisciplinary approach utilizing chemical, ecological, and ecotoxicological data suggested ways to develop efficient and accurate methods to effectively assess and identify the magnitude of genetic changes taking place in any organism including plants and different animal species (Mauderly and Samet, 2009; Pogányova et al., 2019).

GENOTOXICITY BIOMARKERS

Traditionally, biomarkers are defined as biomarkers of exposure, effect, and susceptibility to an individual. A biomarker can be any substance, structure, or process that can be monitored in tissues or fluids and that predicts or influences health; or that assesses the incidence or biological behavior of a disease, but is not a measure of disease, disorder, or health condition itself. Ideally, biomarkers should be accessible (non-invasive), non-destructive, easy, and cost effective to measure. One of the criteria for establishing an association between the exposure and the disease is the biological plausibility. In the context of genotoxicity, biomarkers reflect on to a particular factor associated to carcinogenicity (Mayeux, 2004; Strimbu et al., 2010). Biomarkers are known to contribute significantly in the epidemiological studies of cancer (Ladeira, 2017).

iomarkers offer the opportunity to provide scientific confirmation of osed exposure-disease pathways in human populations, since they can be cited as a result of interaction of the biological system with the environment. he increasing demand for information about health risks derived from exposure to complex mixtures calls for the identification of biomarkers to evaluate genotoxic effects associated with occupational and environmental exposure to chemicals, and other potential sources of damage. An important group of effect of biomarkers is the genotoxicity biomarkers, which have been developed by conducting studies *in vitro* (cells and cell lines), *in vivo* (animals) and *ex vivo* (cells from humans). Cytogenetic biomarkers are the most frequently used endpoints in human biomonitoring studies, and are extensively used to assess the impact of environmental, occupational, and other factors (Tripathi et al., 2015) in genetic instability. Among the wide range of cytogenetic biomarkers, micronuclei in lymphocytes provide a promising approach to assess health risks (Franco et al., 2008; Sharma et al., 2012a; 2012b; DeBord et al., 2015). Out of these, six basic biomarkers are explored by several groups of workers in their studies including (i) the comet assay (ii) micronucleus (MN) and sister chromatid exchange (iii) ^{32}P-postlabeling (iv) the standard molecular end points for DNA damage (v) the traditional cytogenetic end points, chromosome aberrations (CAs), and (vi) urinary 8-hydroxydeoxyguanosine (Mahmoodi et al., 2017).

In order to measure the oxidative DNA damage by most of the currently available methods, one must first isolate DNA. The isolated DNA is then hydrolyzed and the hydrolysate prepared for analysis of oxidized bases. DNA is known to be susceptible to chemical oxidation, the most sensitive base being guanine. The processes of isolation, hydrolysis, and analysis are reported to cause further artifactual oxidation of DNA (especially of guanine residues), raising further the apparent level of base oxidation products and thus invalidating the measurement (Collins et al., 1996; Halliwell, 2000; Guetens et al., 2002).

Toward measuring the oxidative damage of cellular DNA, workers have developed antibody-based methods, which help in visualization of cellular DNA damage. This method is semiquantitative in nature. In this context, the comet assay is commonly applied to directly observe and measure the extent of DNA strand(s) breaks. If a digestion step with DNA repair enzymes is introduced in the protocol, the increased numbers of DNA strand breaks can be used to estimate the level of oxidized DNA bases in the cell. The values so obtained are generally lower than those generated by HPLC analysis of isolated enzymically digested DNA from the same cells (Halliwell, 2000; Fang et al., 2015).

The most commonly used biological materials for studying genotoxic effects in the biomonitoring of the organisms comprise the blood lymphocytes and exfoliated cells, both being easy to sample. Lymphocytes circulate throughout the body, have a reasonably long-life span, and can therefore be damaged in any specific target tissue by a toxic substance. The exfoliated buc cal cells have been effective in showing the genotoxic effects of lifestyle factors such as tobacco smoking, alcohol, medical treatments such as radiotherapy as well occupational and environmental exposure namely exposure to potentially mutagenic and/or carcinogenic chemicals, and in studies of chemoprevention of cancer using antioxidants and the evaluation of malignant transformation of preneoplastic lesions associated with the oral squamous cell carcinoma (Fenech et al., 1999; Majer et al., 2001; Burgaz, 2002; Cavallo et al., 2009).

CYTOKINESIS BLOCKED MICRONUCLEUS (CBMN) ASSAY

Living organisms may be exposed to mutagenic substances that cause cellular damage, which may be induced by chemical, physical, or biological agents that affect DNA, chromosome replication, and gene transcription, which introduce abnormalities leading to onset of cancer and cell death. The CBMN assay is a comprehensive system for measuring DNA damage, cytostasis and cytotoxicity-DNA damage events scored specifically in once-divided binucleated cells. It is a method for assessing DNA damage caused by xenobiotics, allowing detection of effects caused by clastogenic agents (that provoke chromosome breakage) and aneugenic agents (abnormal chromosome segregation associated with loss). Other endpoints that can be measured are nucleoplasmic bridges (NPB), a biomarker of DNA mis-repair and/or telomere end-fusions, and nuclear buds (NBUD), a biomarker of elimination of amplified DNA and/or DNA repair complexes. In summary, it is considered a very reliable assay for the assessment of the genetic damage in the biological cells. This assay system has applications for various purposes including ecotoxicology, nutrition, and radiation sensitivity testing for radiotherapy and cancer risk assessment. It is also useful for evaluation and monitoring of new pharmaceuticals and other drugs (Fenech, 2006, 2007).

COMET ASSAY

The comet assay, also called single-cell gel electrophoresis (SCGE), is a simple and sensitive method for detecting DNA-strand breaks. DNA-strand breaks can originate from the direct modification of DNA by chemical agents or their metabolites; from the processes of DNA excision repair, replication,

.nd recombination; or through apoptosis. A direct breakage of DNA strands takes place when the reactive oxidative species (ROS) interact with DNA. The alkali-labile sites generated by loss of bases in the DNA are converted to strand breaks by alkaline treatment at pH above 13.1 and they are also detected with the comet assay. Comet assay has become one of the standard methods for assessing DNA damage, with a wide range of applications, namely in genotoxicity testing, human biomonitoring, molecular epidemiology, and eco-genotoxicology (monitoring environmental pollution by studying sentinel organisms). This technique is useful in research to study the oxidative damage of DNA as a factor for occurrence of disease, monitoring oxidative stress in animals or human subjects resulting from exercise, or diet, or exposure to environmental agents as well as monitoring DNA damage and repair (Moller et al., 2000; Collins, 2004; Geras et al., 2005; Moller 2005; Collins and Dusinska, 2009).

EXPERIMENTAL MODELS USED FOR STUDY OF GENOTOXICITY

Generation of DNA damage is considered to be an important initial event in carcinogenesis. A considerable battery of assays exists for the detection of different genotoxic effects of compounds in experimental systems, or for investigations of exposure to genotoxic agents in environmental or occupational settings. Some of the tests may have limited use because of their complicated technical setup or because they are applicable to a few cell types. The single cell gel electrophoresis assay system is technically very simple, relatively fast, and cost effective. Using this technique, one can investigate DNA damage in all the mammalian cell types.

While developing various genotoxicity tests in genetic toxicology with many environmental agents, the procedure may encompass both experimental animal models and biomonitoring. The commonly used cytogenetic assays for *in vitro* genotoxicity evaluation are the micronucleus formation (OECD, 2016a) and comet (OECD, 2016b) tests. The micronucleus test is based on the presence of an additional nucleus separated from the main nucleus of a cell and consists of chromosomes or fragment of chromosomes that are not included in the main nucleus during mitosis. The comet assay is a technique for detecting the presence of single strand breaks (SSB) of DNA lesions at alkali-sensitive sites and SSB at sites of incomplete excision repair in mammalian cells (OECD, 2016b). A systemic study on the experimental model(s) used for the assessment of environmental genotoxicity from prokaryotes to eukaryotes have been illustrated below.

BACTERIA

The first study to assess the genetic damage in bacteria treated with 12.5–100 rad of X-rays using comet assay was conducted by Singh et al. (1999). In this study, the neutral comet assay was used for direct (visual) determination of cellular DNA double-strand breaks in the single electrostretched DNA molecule of Escherichia coli JM101. A significant increase in the DNA breaks was induced by a dose as low as 25 rad, which was directly correlated to X-ray dosage. The study supported a hypothesis that strands of the electrostretched human DNA in the comet assay represented individual chromosomes.

PLANT MODELS

Plant bioassays help detect genotoxic contamination in the environment (Maluszynska and Juchimiuk, 2005). Plant systems can provide information about a wide range of genetic damage, including gene mutations and chromosome aberrations. The mitotic cells of plant roots have been used for the detection of clastogenicity of environmental pollutants, especially for *in situ* monitoring of water contaminants. Roots of *Vicia faba* and *Allium cepa* have long been used for assessment of chromosome aberrations (Grant, 1999) and micronucleus (Ma et al., 1995). During the last decade, the comet assay has been extensively applied to plants (leaves, shoot, and roots) to detect DNA damage arising due to chemicals and heavy metals in polluted soil.

LOWER PLANTS

Fungi

Schizosaccharomyces pombe has been used as a model organism to investigate DNA damage due to chlorinated disinfectant, alum, and polymeric coagulant mixture in drinking water samples (Banerjee et al., 2008). The authors observed a significantly higher ($P < 0.001$) DNA damage in chlorinated water ap water) when compared to untreated (negative control) or distilled water oratory control). Hahn and Hock (1999) used mycelia of *Sordaria macros-ra* grown and treated with a variety of DNA-damaging agents directly on garose minigels for assessment of genotoxicity using the comet assay. DNA strand breaks were detected by an increase in the DNA migration from the nucleus. This model allowed the rapid and sensitive detection of DNA damage by a number of chemicals simultaneously.

Algae

Aquatic unicellular plants like algae provide information of potential genotoxicity of the water in which they grow. Being single-celled orgamisms, they can be used as a model for assessment of cDNA damage and monitoring of environmental pollution utilizing comet assay. Unicellular green alga *Chlamydomonas reinhardtii* was used for evaluation of DNA damage due to known genotoxic chemicals. It also demonstrated that oxidative stress was better managed by the algal cells under light than dark conditions (Erbes et al., 1997). The comet assay was found to be useful for evaluating chemically induced DNA damage and repair in *Euglena gracilis*, and responses were more sensitive than those of human lymphocytes under the same treatment conditions (Aoyama et al., 2003). The ease of culturing and handling of *E. gracilis* as well as its sensitivity makes it a useful tool for testing the genotoxicity of chemicals and monitoring environmental pollution. A modified version of the comet assay was used as an alternative technique to assess DNA damage due to UV radiation in *Rhodomonas* sp.(Cryptophyta), a marine unicellular flagellate (Sastre et al., 2001).

Higher plants

V. faba has been widely used for the assessment of DNA damage using comet assay. The strand breaks and abasic (AP) sites in meristematic nuclei of *V. faba* root tips were studied by the neutral and alkaline comet assay (Angelis et al., 2000; Menke et al., 2000). The tobacco and potato plants with increased DNA damage were also found to be severely injured (inhibited growth, distorted leaves), which may be associated with necrotic or apoptotic DNA fragmentation. The major drawback with plant models was the fact that exposure needs to be given in the soil, and it is difficult to say whether the results demonstrate synergies with other chemicals in the soil or non-availability of the toxicant due to its soil binding affinity. Therefore, Vajpayee et al. (2006) used *Bacopa monnieri* L., a wetland plant, as a model for the assessment of ecogenotoxicity using the comet assay.

Animal models

In order to assess safety/toxicity of chemicals/finished products, animal models have long been used. With the advancements in technology, the knockouts and transgenic models have become common to mimic the effects in humans. Comet assay has globally been used for assessment of DNA damage in various animal models.

Lower animals

Tetrahymena thermophila is a unicellular protozoan widely used for genetic studies due to its well characterized genome. Its uniqueness lies in the fact that it has a somatic and a germ nucleus in the same cell. Therefore, it has been validated as a model organism for assessing DNA damage using a modified comet assay protocol standardized with known mutagens such as phenol, hydrogen peroxide, and formaldehyde (Lah et al., 2004). The method was then used for the assessment of genotoxic potential of influent and effluent water samples from a local municipal wastewater treatment plant (Lah et al., 2004). The method provided an excellent, low-level detection of genotoxicants and proved to be a cost effective and reliable tool for genotoxicity screening of wastewater.

Invertebrates

Studies have been carried out on various aquatic (marine and freshwater) and terrestrial invertebrates. The genotoxicity assessment in marine and freshwater invertebrates using the assay has been reviewed (Mitchelmore and Chipman, 1998a; Cotelle and Ferard, 1999; Lee and Steinert, 2003). The cells from hemolymph, embryos, gills, digestive glands, and coelomocytes from mussels (Rank et al., 2005), zebra mussel (Dreissenapolymorpha), clams (Mya arenaria), and polychaetes (Nereis virens) have been used for eco-genotoxicity studies using the comet assay. DNA damage has also been assessed in earthworms (Salagovic et al., 1996; Rajaguru et al., 2003) and the fruit fly e.g. Drosophila species (Bilbao et al., 2002; Mukhopadhyay et al., 2004).

Mussels

Freshwater and marine mussels have been used to study the adverse effect of contaminants in aquatic environment, as they serve as important pollution indicators. The comet assay in mussels can be used to detect a reduction in water quality caused by chemical (Steinert et al., 1998; Frenzilli et al., 2001;

Rank et al., 2005; Jha et al., 2005). *Mytilus edulis* has been widely used for comet assay studies to evaluate DNA-strand breaks in gill and digestive gland nuclei due to polycyclic aromatic hydrocarbons (PAHs) including benzo[α] pyrene (B[α]P; Large et al., 2002) and oil spills with petroleum hydrocarbons (Hamounten et al., 2002). However, the damage returned to normal levels after continued exposure to high dose (20 ppb-exposed diet) of B[α]P for 14 days. Significant levels of inter-individual variability, including seasonal variations in DNA damage, have been reported from some studies in both laboratory and field conditions (Wilson et al., 1998; Frenzilli et al., 2001; Shaw et al., 2000, 2004).

In vitro comet assay has also been used in cells of mussels. Dose–response increases in DNA-strand breakages were recorded in digestive gland cells (Mitchelmore et al., 1998a, 1998b), hemocytes (Rank and Jensen, 2003), and gill cells (Wilson et al., 1998; Rank and Jensen, 2003) of *M. edulis* exposed to both direct (hydrogen peroxide and 3-chloro-4-(dichloromethyl)-5-hydro xy-2[5H]-furanone) and indirect (B[α]P, 1-nitropyrene, nitrofurantoin and N-nitrosodimethylamine) acting genotoxicants.

Coughlan et al. (2002) showed that the comet assay could be used as a tool for the detection of DNA damage in clams (Tapes semidecussatus) as biomonitor organisms for sediments. Significant DNA-strand breaks were observed in cells isolated from haemolymph, gill, and digestive gland from clams exposed to polluted sediment (Coughlan et al., 2002; Hartl et al., 2004). Comet assay was used for the assessment of sperm DNA quality of cryopreserved semen in Pacific oyster (Crassostrea gigas), as it is widely used for artificial fertilization (Gwo et al., 2003). The comet assay detecting DNA-strand breaks has demonstrated that higher basal levels of DNA damage are observed in marine invertebrates; hence, the protocol followed in these animals should be considered for biomonitoring the ecogenotoxicity of a region (Machella et al., 2006).

Earthworm

The comet assay applied to earthworms is a valuable tool for monitoring and detection of genotoxic compounds in terrestrial ecosystems (Salagovic et al., 1996; Zang et al., 2000). As the worms feed on the soil they live in, they are a good indicator of the genotoxic potential of the contaminants present in the soil and thus used as a sentinel species. The *in vitro* exposure of coelomocytes primary cultures to nickel chloride as well as whole animals either in spiked artificial soil water or in spiked cattle manure substrates exhibited increased DNA strand breaks due to the heavy metals (Reinecke and Reinecke, 2004).

Fourie et al. (2007) used five earthworm species (*Amynthas diffringens, Aporrectodea caliginosa, Dendrodrilus rubidus, Eisenia foetida, and*

Microchaetus benhami) to study genotoxicity of sublethal concentration cadmium sulfate. They observed significant level of DNA damage as detect in *E. foetida* followed by *D. rubidus* and *A. caliginosa*. This study showed difference in sensitivity of species present in an environment and its influence on the genotoxicity risk assessment (Fourie et al., 2007).

Drosophila

The simple genetics and developmental biology of *Drosophila melanogaster* has made it the most widely used insect model and has been recommended as an alternate animal model by the European Centre for the Validation of Alternative Methods (ECVAM; Benford et al., 2000).

Some workers have shown that exposure of *D. melanogaster* to cypermethrin, a synthetic pyrethroid, even at very low concentrations (at 0.002 ppm), and leachates of industrial waste caused significant level of DNA damage in the ganglia and anterior midgut of *D. melanogaster* brain (Mukhopadhyay et al., 2004; Siddique et al., 2005b). The results from comet assay have also shown a direct correlation between the concentrations of cisplatin adducts and DNA damage in somatic cells of *D. melanogaster* (García Sar et al, 2008). The *in vitro* studies using Drosophila S2 cells demonstrated that the ectopically expressed DNA glycosylases (dOgg1 and RpS3) reduced the oxidized guanosine (8-OxoG) but contributed to increased DNA degradation due to inhibition of one of the constituents of the DNA repair system (Radyuk et al., 2006). The studies in Drosophila have shown it to be a good alternate to animal model for the assessment of *in vivo* genotoxicity of chemicals using the comet assay.

Vertebrates

Studies of vertebrate species where the comet assay is used include fishes, amphibians, birds, and mammals. The cells (blood, gills, kidneys, and livers) of different fishes, tadpoles, and adult frogs, as well as rodents have been used for assessing *in vivo* and *in vitro* genotoxicity of chemicals. Human biomonitoring has also been carried out employing the comet assay (Dhawan et al., 2009).

Fish

Both freshwater and marine fish species have been employed for studying environmental biomonitoring. These fish species are endemic organisms often as sentinel species inhabiting in a specific aquatic zone, when exposed the chemicals and environmental conditions leading to the onset of their adverse effects.

Environmental biomonitoring to assess the genotoxic potential of river water has been carried out in hepatocytes of chub (*Leuciscus cephalus*) (Winter et al., 2004), erythrocytes of mullet (Mugil sp.), sea catfish (Netuma sp.; de Andrade et al., 2004a, b), bullheads (Ameiurus nebulosus), and carps (Cyprinus carpio; Pandrangi et al., 1995; Buschini et al., 2004). The basal level of DNA damage has been shown to be influenced by various factors, such as temperature of water in erythrocytes of mullet and sea catfish (de Andrade et al., 2004a, b), age, and gender in dab (Limandalimanda; Akcha et al., 2003), and exhaustive exercise in chub (Aniagu et al., 2006).

Amphibians

Comet assay in amphibians has been carried out on adult and larval stages for studying the eco-genotoxicity of aquatic environments, and the results have been analyzed and reviewed by Cotelle and Ferard (1999). The animals chosen for the comet assay act as sensitive bioindicators of aquatic and agricultural ecosystems. Huang et al. (2007) have shown the genotoxicity of petrochemicals in liver and erythrocytes of toad Bufo raddeis. In addition, DNA damage was also found to be positively correlated to the increased concentration of petrochemicals in liver, pointing to the fact that liver is the site for metabolism and may be a good marker for studying genotoxicity of compounds which require metabolic activation. The comet assay has been employed by some workers to study the effect of polyploidy on bleomycin-induced DNA damage as well as repair in *Xenopus laevis* (pseudotetraploid) and *Xenopus tropicalis* (diploid) (Banner et al., 2007).

Birds

There are few studies involving comet assay in birds. Genetic damage due to a mining accident involving heavy metals has been reported in free-living, nestling white storks (Ciconia ciconia), and black kites (Milvus migrans) from southwestern Spain (Baos et al., 2006; Pastor et al., 2001a, b, 2004). However, species specific and intra-species differences in this process have been observed. Frankic et al. (2006) have reported T-2 toxin and deoxynivalenol (DON) induced DNA fragmentation in chicken spleen leukocytes. It was, however, abrogated by dietary nucleotides.

Rodents

Mice and rats have been widely used as animal models for the assessment of *in vivo* genotoxicity of chemicals using the comet assay. The UK Committee on Mutagenicity testing of chemicals in food, consumer products, and environment (COM, 2000) has accepted the *in vivo* comet assay as a test for the assessment of DNA damage. The committee has recommended for follow-up testing of positive *in vitro* cases. In the *in vivo* comet assay, positive results assume the mutagenic potential of any chemical (Brendler Schwaab et al., 2005).

Multiple organs of mouse/rat including brain, blood, kidney, lungs, liver, and bone marrow have been utilized for the comprehensive understanding of the systemic genotoxicity of chemicals (Meng et al., 2004; Patel et al., 2006; Sasaki et al., 2000; Sekihashi et al., 2002). A comprehensive data on chemicals representing different classes, e.g., PAHs, alkylating compounds, nitroso compounds, food additives, etc., that caused DNA-strand breaks in various organs of mice as has been compiled by Sasaki et al. (2000, 2002). The *in vivo* comet assay in rodents is an important test model for genotoxicity studies. Many rodent carcinogens are also human carcinogens. Hence this model not only provides an insight into the genotoxicity of human carcinogens but also is suited for studying their underlying mechanisms.

Humans

Comet assay is a valuable method for detection of occupational and environmental exposures to genotoxicants in humans and can be used as a tool in risk assessment for hazardous chemicals (Albertini et al., 2000; Moller 2005, 2006a; Dusinska and Collins, 2008; Dhawan et al., 2009). As an alternate of the cytogenetic assays in early genotoxicity/neurotoxicity/photogenotoxi city screening of several drug candidates, the *in vitro* comet assay has been suggested (Witte et al., 2007). Certain factors like age, diet, lifestyle (alcohol and smoking), as well as diseases have been shown to influence the comet assay parameters; and for interpretation of responses, these factors need to be accounted for during monitoring human genotoxicity (Moller et al., 2000; Anderson, 2001). Human biomonitoring studies using the comet assay provide an efficient tool for measuring human exposure to genotoxicants, thus helping in risk assessment and hazard identification.

The results of a large Japanese study have demonstrated that the discrimination between carcinogens and non-carcinogens appears to be similar between the comet assay and alkaline elution. This suggests that the comet assay is a reliable genotoxicity test in animal experimental systems. In the biomonitoring studies, some authors have investigated the effect of common exposures and lifestyle factors on the level of oxidative DNA damage in mononuclear blood cells of humans. The comet assay indicated that DNA damage was abundant in mammalian cells, though it was affected by lifestyle and many environmental exposures, including diet, exercise, hypoxia, and sunlight (Dhawan et al., 2009).

EXPERIMENTAL MODELS FOR STUDY OF DNA PROTECTIVE POTENTIAL

The results from many studies have primarily focused on the use of antioxidant vitamins such as vitamins A, C, and E, the carotenoids, juices of fruits specially from grapes, kiwi, beverages such as soy milk, tea, and coffee), vegetables including tomato products, berries, Brussels sprouts, and several other components of the human diet (coenzyme Q10) containing polyunsaturated fatty acids. On the basis of the results of these studies it was possible to identify dietary compounds which were highly active such as gallic acid from Carob fruit. At present, strong efforts are being made to elucidate whether the different parameters of oxidative DNA-damage correlate with life span, cancer, and other age-related diseases. The new techniques are highly useful in providing valuable information if the dietary components may exert antioxidant effects in humans and can be used to identify individual protective compounds and also to develop nutritional strategies to reduce the adverse health effects of ROS (Hoelzl et al., 2005). The antioxidant potential of indolinic and quinolinic nitroxide radicals (Villarini et al., 1998), tannins (Fedeli et al., 2004), and low concentrations (<10 μM) of diaryl tellurides and ebselen, an organoselenium compound (Tiano et al., 2000), and their protective effect in oxidative DNA damage has been studied in nucleated trout (*Oncorhynchus mykiss*) erythrocytes. The results obtained from the experiments conducted by some workers have suggested that antioxidant supplementation can improve the sensitivity of the comet assay by lowering the baseline damage in untreated animals (Wilson et al., 1998). In another such study, the authors aimed to investigate intra- and inter-individual differences in buccal cell DNA damage (as strand breaks), the effect of *in vitro* exposure to both a standard oxidant challenge and antioxidant treatment, as well as *in situ* exposure to an antioxidant-rich beverage and supplementation-related effects using a carotenoid-rich food. They observed that the exposure of buccal cell *in situ* (i.e. in the mouth) to antioxidant-rich green tea led to an acute decrease in basal DNA-strand breaks. In a controlled human intervention trial, buccal cells from 14 subjects after 28 days' supplement with a carotenoid-rich berry (Fructus barbarum L.) showed a small but sta tically significant ($P < 0.05$) decrease in DNA strand breaks. Their data ar indicated that this buccal cell comet assay was a feasible and potentially usefu alternative tool to the usual lymphocyte model in human biomonitoring and nutritional work (Szeto et al., 2005).

In another study, the antigenotoxic potential of methyl methanesulfo-nate (MMS) have been evaluated against genotoxic effects of (+)-usnic acid (UA) using the micronucleus and comet assays in V79 cell cultures (treated with 15, 30, 60, and 120 mg/mL UA) and Swiss mice (treated with UA doses of 25, 50, 100, and 200 mg/kg body weight). The same concentrations of UA combined for evaluation of antigenotoxicity. Combined administra-tion of UA with MMS significantly reduced the frequencies of micronuclei formation and DNA damage *in vitro* and *in vivo* when compared to the treatment with MMS alone. Although the mechanisms underlying the pro-tective effect of UA are not completely understood, the antioxidant activity of this metabolite may explain its protective effect against MMS-induced genotoxicity (Leandro et al., 2013). Jothy et al. (2013) have investigated the protective effect of *P. longifolia* on DNA damage-induced by hydroxyl radicals using a plasmid, and *Allium cepa* extract in the comet assay. In the presence of •OH radicals, the DNA in supercoil got nicked into open circular form. The product contained single-stranded cleavage of supercoil DNA and quantified as fragmented separate bands on agarose gel in plasmid relation assay. In the plasmid relation and comet assays, the *P. longifolia* leaf extract exhibited strong inhibitory effects against H_2O_2-mediated DNA damage. A dose-dependent increase of chromosome aberrations was also be observed in the *Allium cepa* assay. The results of *Allium cepa* assay confirmed that the methanol extracts of *P. longifolia* exerted no significant genotoxic or mitodepressive effects at 100 µg/mL. This study demonstrated that *P. longifolia* leaf extract exhibited a beneficial effect against oxidative DNA damage (Jothy et al., 2013).

CONCLUSION

The reports available as on date indicated that both the xenobiotics and some natural compounds present in the environment have abilities to cause DNA damage. Many in vitro, in vivo, and ex vivo experimental models have been explored and utilized to study DNA damage. Several methods have been developed to assess the extent of DNA damage, the comet assay being most _ominantly used. Some plant extracts have the potential to protect the _A from being damaged by the chemicals, radiations, or natural toxicants. _owever, extensive studies are needed to generate more information on the _ubject so as to keep the living systems safe from the ill effects of environmental pollutants.

REFERENCES

Akcha, F., Vincent Hubert, F., Pfhol-Leszkowicz, A. 2003. Potential value of the Comet assay and DNA adduct measurement in dab (*Limanda limanda*) for assessment of in situ exposure to genotoxic compounds. *Mutation Research* 534(1–2); 21–32.

Angelis, K.J., McGuffie, M., Menke, M., Schubert, I. 2000. Adaptation to alkylation damage in DNA measured by the Comet assay. *Environmental and Molecular Mutagenesis* 36(2): 146–150.

Aniagu, S.O., Day, N., Chipman, J.K., Taylor, E.W., Butler, P.J. 2006. Winter, M.J. Does exhaustive exercise result in oxidative stress and associated DNA damage in the chub (*Leuciscus cephalus*)? *Environmental and Molecular Mutagenesis* 47(8); 616–623.

Aoyama, K., Iwahori, K., Miyata N. 2003. Application of Euglena gracilis cells to comet assay: evaluation of DNA damage and repair. *Mutation Research* July 8; 538(1–2); 155–162.

Banerjee, P., Talapatra, S.N., Mandal, N. et al. 2008. Genotoxicity study with special reference to DNA damage by Comet assay in fission yeast, Schizosaccharomyces pombe exposed to drinking water. *Food and Chemical Toxicology* 46(1); 402–407.

Bilbao, C., Ferreiro, J.A., Comendador, M.A., Sierra, L.M. 2002. Influence of mus201 and mus308 mutations of *Drosophila melanogaster* on the genotoxicity of model chemicals in somatic cells *in vivo* measured with the comet assay. *Mutation Research*, 503; 11–19.

Burgaz, S., Erdem, O., Cakmak, G. et al. 2002. Cytogenetic analysis of buccal cells from shoe-workers and pathology and anatomy laboratory workers exposed to n-hexane, toluene, methyl ethyl ketone and formaldehyde. *Biomarkers* 7(2); 151–161.

Cavallo, D., Ursini, C.L., Rondinone, B., Iavicoli, S. 2009. Evaluation of a suitable DNA damage biomarker for human biomonitoring of exposed workers. *Environmental and Molecular Mutagenesis* 50(9); 781–790.

Chaudhari, R., Saxena, K.K. 2016. Genotoxicological assessment of pyrethroid insecticide bioallethrin in freshwater fish Channa punctatus. *RJLBPCS* 2; 55–62.

Collins, A.R., Dusinska, M.; 2009. Applications of the comet assay in human biomonitoring. In: *The Comet Assay in Toxicology*, Dhawan, A. and Anderson, D. (eds.), Royal Society of Chemistry, Cambridge, 201–202.

Collins, A.R., Dusinska, M., Gedik, C.M., Stetina, R. 1996. Oxidative damage to DNA: Do we have a reliable biomarker? *Environ Health Perspect* 104(supplement 3); 465–469.

Collins, A.R. 2004. The Comet assay for DNA damage and repair principles, applications, and limitations. *Molecular Biotechnology* 26; 249–260.

Cotelle, S., Férard, J.F. 1999. Comet assay in genetic ecotoxicology: A review, *Environmental and Molecular Mutagenesis* 34; 246–255.

de Andrade, V.M., de Freitas, T.R., da Silva, J. 2004a. Comet assay using mullet (*Mugil* sp.) and sea catfish (*Netuma* sp.) erythrocytes for the detection of genotoxic pollutants in aquatic environment. *Mutation Research* 560(1); 57–67.

de Andrade, V.M., da Silva, J., da Silva, F.R., Heuser, V.D., Dias, J.F., Yoneama, M.L., de Freitas, T.R.O. 2004b. Fish as bioindicators to assess the effects of pollution in two southern Brazilian rivers using the Comet assay and micronucleus test. *Environmental and Molecular Mutagenesis* 44(5); 459–468. https://doi.org/10.1002/em.20070

De Bord, D.G., Burgoon, L., Edwards, S.W.. et al. 2015. Systems biology and biomarkers of early effects for occupational exposure limit setting. *Journal of Occupational and Environmental Hygiene* 1(supplement 1); S41–S54.

Dhawan, A., Bajpayee, M., Parmar, D. 2009. Comet assay: a reliable tool for the assessment of DNA damage in different models. *Cell Biology and Toxicology* 25; 5–32.

Erbes, M., Wessler, A., Obst, U., Wild, A. 1997. Detection of primary DNA damage in Chlamydomonas reinhardtii by means of modified microgel electrophoresis. *Environmental and Molecular Mutagenesis* 30(4); 448–458.

Fang, L., Neutzner, A., Turtschi, S. et al. 2015. Comet assay as an indirect measure of systemic oxidative stress. *Journal of Visualized Experiments* 99(99); e52763.

Fenech, M. 2006. Cytokinesis-block micronucleus assay evolves into a 'cytome' assay of chromosomal instability, mitotic dysfunction and cell death. *Mutation Research* 600(1–2); 58–66.

Fenech, M. 2007. Cytokinesis-block micronucleus cytome assay. *Nature Protocols* 2(5); 1084–1104.

Fenech, M., Crott, J., Turner, J., Brown, S. 1999. Necrosis, apoptosis, cytostasis and DNA damage in human lymphocytes measured simultaneously within the cytokinesis-block micronucleus assay: Description of the method and results for hydrogen peroxide. *Mutagenesis* 14(6): 605–612.

Fourie, F., Reinecke, S.A., Reinecke, A.J. 2007. The determination of earthworm species sensitivity differences to cadmium genotoxicity using the Comet assay. *Eco toxicology and Environmental Safety* 67(3); 361–368.

Franco, S.S., Nardocci, A.C., Gunther, W.M.R. 2008. PAH biomarkers for human health risk assessment: A review of the state-of-the-art. *Cad Saúdepubl* 24(supplement 4); a569–s580.

Frenzilli, G., Nigro, M., Scarcelli, V., Gorbi, S., Regoli, F. 2001. DNA integrity and total oxyradical scavenging capacity in the Mediterranean mussel, Mytilus galloprovincialis: A field study in a highly eutrophicated coastal lagoon. *Aquatic Toxicology* 53(1); 19–32.

García Sar, D., Montes-Bayón, M., Aguado Ortiz, L., Blanco-González, E., Sierra, L.M., Sanz-Medel A. 2008. In vivo detection of DNA adducts induced by cisplatin using capillary HPLC-ICP-MS and their correlation with genotoxic damage in Drosophila melanogaster. *Analytical and Bioanalytical Chemistry* 390(1); 37–44.

Geraskina, S.A., Kimb, J.K., Oudalovaa, A.A. et al. 2005. Bio-monitoring the geno-toxicity of populations of Scots pinein the vicinity of a radioactive waste storage facility. *Mutation Research* 583(1); 55–66.

Grant, W.F. 1999. Higher plant assays for the detection of chromosomal aberrations and gene mutations-a brief historical background on their use for screening and monitoring environmental chemicals. *Mutation Research* May 19; 426(2); 107–112. doi: 10.1016/s0027-5107(99)00050-0.

Guetens, G., Boeck, G.D., Highley, M. et al. 2002. Oxidative DNA damage: Biological significance and methods of analysis. *Critical Reviews in Clinical Laboratory Sciences* 39(4–5); 331–457.

Halliwell, B. 2000. Why and how should we measure oxidative DNA damage in nutritional studies? How far have we come? *American Journal of Clinical Nutrition* 72(5); 1082–1087.

Hahn, A., Hock, B. 1999. Assessment of DNA damage in filamentous fungi by single cell gel electrophoresis, Comet assay. *Environmental Toxicology and Chemistry* 18; 1421–1424.

Hamouten, D., Payne, J.F., Rahimtula, A., Lee, K. 2002. Use of the Comet assay to assess DNA damage in hemocytes and digestive gland cells of mussels and clams exposed to water contaminated with petroleum hydrocarbons. *Marine Environmental Research*54(3–5); 471–474.

Hartwig, A., Arand, M., Epe, B. et al. 2020. Mode of action-based risk assessment of genotoxic carcinogens [published correction appears in. *Archives of Toxicology* 94(6); 1787–1877.

Hernandez, L.M. and Blazer, D.G.. *Genes, Behavior, and the Social Environment.* Institute of Medicine (US) Committee on Assessing Interactions Among Social, Behavioral, and Genetic Factors in Health.

Hoelzl, C., Bichler, J., Ferk, F. et al. 2005. Methods for the detection of antioxidants which prevent age related diseases: A critical review with particular emphasis on human intervention studies. *Journal of Physiology Pharmacology* 2(supplement 2); 49–64.

Jha, A.N., Dogra, Y., Turner, A., Millward, G.E. 2005. Impact of low doses of tritium on the marine mussel, Mytilus edulis: genotoxic effects and tissue-specific bio-concentration. *Mutation Research* 586(1); 47–57.

Jothy, S.L., Chen, Y., Kanwar, J.R., Sasidharan, S. 2013. Evaluation of the geno-toxic potential against H_2O_2 radical-mediated DNA damage and acute oral toxicity of standardized extract of *Polyalthialongifolia*Leaf. *Evidence-Based Complementary and Alternative Medicine* 2013; 1–13.

Ladeira, C., Smajdova, L. 2017. The use of genotoxicity biomarkers in molecular epidemiology: Applications in environmental, occupational and dietary studies. *AIMS Genetics* 4(3):166–191.

Lah, B., Malovrh, S., Narat, M., Cepeljnik, T., Marinsek-Logar R. 2004. Detection and quantification of genotoxicity in wastewater-treated *Tetrahymena thermophila* using the Comet assay. *Environmental Toxicology* 19(6): 545–553.

Large, A.T., Shaw, J.P., Peters, L.D., McIntosh, A.D., Webster, L., Mally, A., et al. Different levels of mussel (*Mytilus edulis*) DNA strand breaks following chronic field and acute laboratory exposure to polycyclic aromatic hydrocarbons. *Marine Environmental Research* 54(3–5); 493–497.

Lee, R.F., Steinert, S. 2003. Use of the single cell gel electrophoresis/comet assedetecting DNA damage in aquatic (Marine and freshwater) animals. *Mut Research/Fundamental and Molecular Mechanisms of Mutagenesis.* 544 43–64. DOI: 10.1016/S1383-5742(03)00017-6

Leandro, L.F., Munari, C.C., Sato, V.L.F.L. et al. 2013. Assessment of the genotoxicit and antigenotoxicity of (+)-usnic acid in V79 cells and Swiss mice by the micronucleus and comet assays. *Mutation Research* 753(2); 101–106.

Ma, T.H., Xu, Z., Xu, C., McConnell, H., Vailterra Rabago, E., Arreola, A.G., Zhang, H. 1995. The improved Allium/Vicia root tip micronucleous assay for clastogenicity of environmental pollutants. *Mutation Research* 334; 185–195.

Mahmoodi, M., Soleyman-Jahi, S., Zendehdel, K. et al. 2017. Chromosomal aberrations, sister chromatid exchanges, and micronuclei in lymphocytes of oncology department personnel handling anti-neoplastic drugs. *Drug and Chemical Toxicology* 40(2); 235–240.

Majer, B.J., Laky, B., Knasmüller, S., Kassie, F. 2001. Use of the micronucleus assay with exfoliated epithelial cells as a biomarker for monitoring individuals at elevated risk of genetic damage and in chemoprevention trials. *Mutation Research* 489(2–3); 147–172.

Maluszynska, J., Juchimiuk J. 2005. Plant genotoxicity: A molecular cytogenetic approach in plant bioassays. *Arh Hig Rada Toksikol.* Jun, 56(2); 177–84. PMID: 15968834.

Mauderly, J.L., Samet, J.M. 2009. Is there evidence for synergy among air pollutants in causing health effects? *Environmental Health Perspectives* 117(1); 1–6.

Mayeux, R. 2004. Biomarkers: Potential uses and limitations. *Neurorx* 1(2); 182–188.

Menke, M., Meister, A., Schubert, I. 2000. N-Methyl-N-nitrosourea-induced DNA damage detected by the Comet assay in *Vicia faba* nuclei during all interphase stages is not restricted to chromatid aberration hot spots. *Mutagenesis.* 15(6); 503–506.

Mitchelmore, C.L., Chipman, J.K. 1998a. DNA strand breakage in aquatic organisms and the potential value of the Comet assay in environmental monitoring. *Mutation Research* 399; 135–147.

Mitchelmore, C.L., Birmelin, C., Livingstone, D.R., Chipman, J.K. 1998b. Detection of DNA strand breaks in isolated mussel (*Mytilus edulis* L.) digestive gland cells using the "Comet" assay. *Ecotoxicology and Environmental Safety* 41(1); 51–58.

Møller, P. 2005. Genotoxicity of environmental agents assessed by the alkaline comet assay. *Basic Clinical and Pharmacology and Toxicology* 1(supplement 1), 1–42.

Moller, P., Knudsen, L.E., Loft, S., Wallin, H. 2000. The comet assay as a rapid test in biomonitoring occupational exposure to DNA-damaging agents and effect of confounding factors. *Cancer Epidemiology Biomarkers & Prevention* 9(10); 1005–1015.

Mukhopadhyay, Indranil, Kar Chowdhuri, N, Bajpayee, Mahima, Dhawan, Alok. 2004. Evaluation of *in vivo* genotoxicity of cypermethrin in *Drosophila melanogaster* using the alkaline Comet assay. Mutagenesis March, 19 (2); 85–90.

Pandrangi, R., Petras, M., Ralph, S., Vrzoc, M. 1995. Alkaline single cell gel (comet) assay and genotoxicity monitoring using bullheads and carp. *Environmental an d Molecular Mutagenesis* 26; 345–356.

ιyova, A., Mičieta, K., Dušička, J. 2019. Genotoxic assessment of selected native plants to deferentially exposed urban ecosystems. *Environmental Science and Pollution Research International* 26(9); 9055–9064.

.ank, J., Jensen, K., Jespersen, P.H. 2005. Monitoring DNA damage in indigenous blue mussels (*Mytilus edulis*) sampled from coastal sites in Denmark. *Mutation Research* 585(1–2); 33–42.

Radyuk, S.N., Michalak,K., Rebrin, I., Sohal, R.S., Orr, W.C. 2006. Effects of ectopic expression of Drosophila DNA glycosylases dOgg1 and RpS3 in mitochondria. *Free Radical Biology and Medicine Free Radic Biol Med*.41(5); 757–764.

Rajaguru, P., Kalpana, R., Hema, A., Suba, S., Baskarasethupathi, B., Kumar, P.A, et al. 2001. Genotoxicity of some sulfur dyes on tadpoles (*Rana hexadactyla*) measured using the Comet assay. *Environmental and Molecular Mutagenesis* 38(4); 316–322.

Rank, J., Jensen, K. 2003. Comet assay on gill cells and hemocytes from the blue mussel *Mytilus edulis*. *Ecotoxicology and Environmental Safety* 54(3); 323–329.

Reinecke, S.A., Reinecke, A.J. 2004. The Comet assay as biomarker of heavy metal genotoxicity in earthworms. *Archives of Environmental Contamination and Toxicology* 46(2); 208–215.

Rhind, S.M. 2009. Anthropogenic pollutants: A threat to ecosystem sustainability? *Philosophical Transactions of the Royal Society London B Biological Sciences* 364(1534); 3391–3401.

Salagovic, J., Gilles, J., Verschaeve, L., Kalina, I. 1996. The Comet assay for the detection of genotoxic damage in the earthworms: a promising tool for assessing the biological hazards of polluted sites. *Folia Biol (Praha)*. 42(1–2): 17–21.

Sastre, M.P., Vernet, M., Steinert, S. 2001. Single-cell gel/comet assay applied to the analysis of UV radiation-induced DNA damage in Rhodomonas sp. (Cryptophyta). *Photochemistry and Photobiology* July 74(1); 55–60.

Sharma, R.K., Rai, D.K., Sharma, B. 2012b. *In vitro* carbofuran induced micronucleus formation in human blood lymphocytes. *Cellular and Molecular Biology (Noisy-Le-Grand) France)* 58(1); 128–133.

Sharma, R.K., Sharma, B. 2012a. *In vitro* carbofuran induced genotoxicity inhuman lymphocytes and its mitigation by vitamins C and E. *Disease Markers* 32(3); 153–163.

Shaw, J.P., Large, A.T., Chipman, J.K., Livingstone, D.R., Peters L.D. 2000. Seasonal variation in mussel Mytilus edulis digestive gland cytochrome P4501A- and 2E-immunoidentified protein levels and DNA strand breaks (Comet assay). *Marine Environmental Research* 50(1–5); 405–409.

Shaw, J.P., Large, A.T., Donkin, P., Evans, S.V., Staff, F.J., Livingstone, D.R., et al. 2004. Seasonal variation in cytochrome P450 immunopositive protein levels, lipid peroxidation and genetic toxicity in digestive gland of the mussel Mytilus edulis. *Aquatic Toxicology* 67(4); 325–336.

Singh, N.P., Stephens, R.E., Singh, H., Lai, H. 1999. Visual quantification of DNA double-strand breaks in bacteria. *Mutation Research* 429; 159–168.

Steinert, S.A., Streib-Montee, R., Leather, J.M., Chadwick, D.B. 1998. DNA damage in mussels at sites in San Diego Bay. *Mutation Research* 399; 65–85.

Strimbu, K., Tavel, J.A. 2010. What are biomarkers? *Current Opinion HIV AIDS* 5(6); 463–466.

Szeto, Y.T., Benzie, I.F.F., Collins, A.R. *et al.* 2005. A buccal cell model comet assay: Development and evaluation for human biomonitoring and nutritional studies. *Mutation Research* 578(1–2); 371–381.

Tiano, L., Fedeli, D., Santroni, A.M., Villarini, M., Engman, L., Falcioni, G. 2000. Effect of three diaryl tellurides, and an organoselenium compound in trout erythrocytes exposed to oxidative stress in vitro. *Mutation Research* 464(2); 269–277.

Tripathi, R., Jaiswal, N., Sharma, B., Malhotra, S.K. 2015. Helminth infections mediated DNA damage: Mechanisms and consequences. *Single Cell Biology* 4; 117. Washington (DC): National Academies Press (Us); 2006. ISBN-10: 0-309-10196-4.

Vajpayee, P., Dhawan, A., Shanker, R. 2006. Evaluation of the alkaline Comet assay conducted with the wetlands plant Bacopa monnieri L. as a model for ecogenotoxicity assessment. *Environmental and Molecular Mutagenesis* Aug 47(7); 483–489.

Vassilev, S.V., Menendez, V.-R. 2005. Phase-mineral and chemical composition of coal fly ashes as a basis for their multicomponent utilization. 4. Characterization of heavy concentrates and improved fly ash residues. *Fuel* 84(7-8); 973–991.

Wilson, J.T., Pascoe, P.L., Parry, J.M., Dixon, D.R. 1998. Evaluation of the Comet assay as a method for the detection of DNA damage in the cells of a marine invertebrate, *Mytilus Edulis* L. (Mollusca: Pelecypoda). *Mutat Res.* 399(1):87–95.

Winter, M.J., Day, N., Hayes, R.A., Taylor, E.W., Butler, P.J., Chipman, J.K. 2004. DNA strand breaks and adducts determined in feral and caged chub (*Leuciscus cephalus*) exposed to rivers exhibiting variable water quality around Birmingham, UK. *Mutation Research* 552(1–2); 163–75.

Zang, Y., Zhong, Y., Luo, Y., Kong, Z.M. 2000. Genotoxicity of two novel pesticides for the earthworm, Eisenia fetida. *Environmental Pollution* 108(2); 271–8.

DNA damage

Overview,
conclusion, and
future perspective

10

OVERVIEW

DNA is the main genetic material in living organisms. Apart from few viruses using RNA as genetic material, all other organisms' genetic material is DNA, which contains all the genetic information needed for the growth, development, and reproduction of organism. DNA is the polymer of deoxyribonucleoside monophosphate connected together by phosphodiester bonds. DNA has two strands of polynucleotide chains, which is a double helix structure with hydrogen bonds between the complementary base pairs. The DNA damage and its effect on living organisms could be significant research areas in life sciences, medicine, and chemistry. In living organisms, DNA is very stable or conservative, and its primary structure is hardly changed during various vital processes. It is the key cellular biomolecule responsible for transferring characteristics of an individual from one generation to another (Chargaff, 1950; Watson and Crick, 1953).

DNA in each living cell serves as the repository of genetic information. The integrity and stability of DNA are highly essential to regulate activities of life of any living systems. The DNA molecules are, however, not inert. It is negatively charged because of the presence of phosphate group in the nucleotides projected outside the surface of helices. It acts as a strong nucleophile. Those chemicals which are strong electrophiles may rapidly interact with DNA and can make adduct, which may generate

~utations or damage to DNA, if the damage stay unrepaired. It is a mac-~omolecule which is always prone to be adversely affected by several environmental factors resulting in DNA damage and onset of several diseases. For instance, DNA damage caused by UV-radiation present in sunlight may lead to skin cancer, whereas smoke from tobacco-mediated DNA damage may induce lung cancer. Another major factor for DNA damage is by the free radicals produced due to use of xenobiotics, which create imbalance in redox system and generate oxidative stress. According to an estimate (Lodish et al., 2004; Clancy, 2008), an individual cell receives insults to its DNA up to one million times a one day.

Most of the DNA damages are restored by an inherent strong repair system in any organism, and the unrepaired DNA damage remains very little or almost nil. But just the very small unrepaired DNA damage may lead to a dramatically significant adverse influence on the organism (Liu Dianfeng, 2006; Hoeijmakers, 2001; Jun, 2010). If this damage occurs in the double-strand of the normal DNA, the structure of the DNA and its function maybe changed. If the damage occurs in the replication process of the DNA, it will lead to base mismatch, deletion, and some other damages. And it will generate erroneously encoded RNA and consequently affect the DNA translation, if the damage occurs in the process of transcription. Gene mutation caused by DNA damage may lead to aging of the organism, cancer, or some other genetic diseases. However, not all gene mutations are harmful to organism. Some of the gene mutations beneficial to organism may be retained, and eventually may lead to biological evolution. DNA damage is thus one of the important means for the biological evolution. Artificial DNA damage or artificially induced mutation is also an important method for the treatment of cancer or genetic diseases. With the advent of genetic engineering techniques, an artificially induced mutation has increasingly been used to improve or even create new species (Brown, 2002; Chatterjee and Walker, 2017).

CAUSES OF DNA DAMAGE

Causes of DNA damage are varied; they could be physical, chemical, or biological factors. DNA damage can be caused by physical factors such as high temperature (heat stress), and radiations (ultraviolet rays and other ionizing radiations). The ultraviolet (UV) rays can cause thymine dimerization to yield thymine dimmers. The formation of thymine dimmers can cause a deformation of the DNA's double helix structure and consequently affect DNA unwinding and other processes, and even lead to the cessation of DNA replication and transcription. In addition, ultraviolet rays can also cause the cross-linking of DNA double strand and DNA with protein. Most of the DNA damages caused by ionizing radiation is not direct damage, but firstly induce a large number of free radicals in an organism, and then these free radicals usually lead to various types of DNA damage. Due to the break of phosphodiester bond caused by ionizing radiation, DNA double-strand breaks are induced (Gros et al., 2002; Cannan and Pederson, 2016). This process associates with the mechanism of ultraviolet-mediated sterilization which is linked to the damage of bacterial DNA, thereby disrupting the replication process of its DNA. The radiation therapy of cancer is also using ionizing radiation to damage the DNA of specific cancer cells, to stop its division and finally to kill them (Rastogi et al., 2010; Ramasamy et al., 2017).

There are many chemical and biological factors that can lead to DNA damage. These agents are known to induce production of excess of free radical species such as $O._2$, H_2O_2, and other reactive oxygen species spontaneously during biological metabolism. Also, the exposure of any living system to heavy metals, medicine, and pesticides and their metabolites, which enter the body through various routes, can lead to DNA damage. Some of the heavy metals have no direct impact to cause damage to DNA, but they play a key role in inducing production of the oxidative chemical species which cause damage of DNA (Phaniendra et al., 2015; Pizzino et al., 2017). Base molecular isomerization can lead to DNA damage. Isomerization can occur on cytosine (C), thymine (T), adenine (A) and guanine (G), namely all of the DNA bases. The isomerization will enable the position of hydrogen bonds between base pairs changed, then result in the bases-mismatched in the replication process, and bring about DNA damage, such as adenine pairing with cytosine (A-T), thymine pairing with guanine (T-G), and so on. Among the four bases of any DNA molecule, cytosine, adenine, and guanine bases have exocyclic-amino (-NH2) group (Brown, 2002; Kino et al., 2017).

Cross-linking of DNA can cause serious DNA damage which can stop the unwinding, the replication, and transcription process of DNA or even lead to the death of cells. In fact, for the therapeutic purposes, some anticancer drugs are just making use of their cross-link reaction with DNA to inhibit the growth, impact division of cancer cells, and finally kill them. Purine bases or pyrimidine bases can be removed by hydrolysis. Base alkylation reaction can also help in the shedding of the bases (Deans, 2011).

The alkylation agent is a reactant that introduces replacement of a hydrogen atom by an alkyl group in the negatively charged DNA or negatively charged groups present on proteins. Some of the alkylation substances are alkylsulfonates (busulfan: ethyl methane sulphonate (EMS), ethyl ethane sulphonate (EES)), cisplatin, ethyleneimines (thiotepa), nitrogen mustards (chlorambucil and cyclophosphamide), nitrosoureas (carmustine, lomustine, and semustine), and triazines (dacarbazine). The alkylating agents belong to the first and foremost group of compounds found to be utilized in the chemotherapy of cancer (Colvin et al., 2005). Most of the alkylating chemicals are monofunctional methylating compounds such as temozolomide [TMZ], -methyl--nitro--nitrosoguanidine [MNNG], and dacarbazine. The alkylating agents which are bifunctional in nature include carmustine [BCNU], chloroethylating agents (e.g., nimustine [ACNU], nitrogen mustards (e.g., chlorambucil and cyclophosphamide), or lomustine [CCNU], and fotemustine (Kondo et al., 2010).

The cytotoxicity caused by alkylating agents follows DNA damage, which is strongly and immediately repaired by suitable DNA repair pathways following different repair mechanisms. It is known that the simple methylating agents commonly form DNA adducts at its N- and O-atoms. The alkylation at N-atom of DNA viz. N-methylations is removed by the process following base excision repair (BER) mechanism, which is also known as the nucleotide excision repair (NER). The alkylation at O-atom of DNA results in the formation of O^6-methylguanine (MeG). This DNA adduct is mutagenic and cytotoxic in nature. It may cause DNA damage. It is repaired by an enzyme called O^6-methylguanine-DNA methyltransferase. The O^6-methylguanine mispairs with deoxythimidine. It can cause DNA damage. This $O6$ MeG:T mispair is therefore recognized by another DNA damage repair system called mismatch repair system (MMR). If the MMR system fails to repair $O6MeG/T$ mispair, it may finally lead to the double-strand breaks (BSB). The bifunctional alkylating chemicals are known to form interstrand cross-links (ICLs). The ICLs are highly complex and cytotoxic in nature. The repair of ICLs is a complex process. They are repaired by NER factors comprising endnuclease xeroderma pigmentosum complementation group Fexcision repair cross-complementing rodent repair deficiency complementation group 1, Fanconi anemia repair, and homologous recombination (Kondo et al., 2010).

DNA alkylation and DNA repair mechanisms are illustrated in the following figure (Figure 10.1).

The base analogues such as 5-bromouracil and 2-aminopurine have been used as a mutagen for creating gene mutation in genetic engineering. They mutate DNA when the above-mentioned analogs are incorporated into DNA undergoing replication (Ang et al., 2016).

The DNA intercalating agents such as berberine, ethidiumbromide, proflavine, daunomycin, doxorubicin, and thalidomide can cause damage to DNA through insertion between the two stacking planar base pairs of DNA.

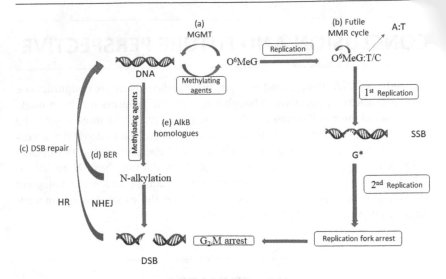

FIGURE 10.1. Pathways showing DNA damage by alkylating agents. (a) The enzyme O^6-methylguanine-DNA methyltransferase (MGMT) eliminates in one step the methyl adduct from O^6MeG. If it is not repaired, the mismatched pairs such as O^6MeG:C or O^6MeG:T mismatch are formed during the course of replication. In the next round of The replication process in its next round converts O^6MeG:T pairs into A:T base pair through transition mutation. (b) The mismatch pair like O^6MeG:T and a pair like O^6MeG:C are acted upon by a mismatch repair (MMR) system. The MMR system generates a double-strand break (DSB) by generating in DNA a single-strand break (SSB), and introducing a pause in the replication process. It has been observed that the O^6MeG:T/C cannot arrest the cell cycle at the first checkpoint of G2/M DNA damage. During the second cell cycle, however, the arrest at G2/M checkpoint takes place. (c) The DSBs repair, on the other hand, occurs via two different processes such as a homologous recombination (HR) and a non-homologous end-joining (NHEJ). The repair of N-alkylation in DNA is performed by either (d) BER, or (e) the AlkB homologues. In case the N-alkylations are not repaired, the DSBs may occur (Kondo et al., 2010).

Peroxides and free radicals can open the ring of DNA bases. Several studies have shown that hydroxyl radicals and superoxide free radicals produced by automobile exhaust, cooking smoke, or cigarette smoke can react with guanine, generating 8-hydroxy-deoxyguanosine. Nowadays, 8-hydroxy-deoxyguanosine has become a recognized biomarker compound of the oxidative damage of DNA. Some other compounds can also modify directly the bases or other position of DNA strand (Mukherjee and Sasikala, 2013).

CONCLUSION AND FUTURE PERSPECTIVE

The study of DNA damage and its repair mechanisms is of great significance from biomedical perspectives. Though a remarkable progress has been made to understand the mechanisms of DNA damage and repair, lot more is yet to be explored. Through direct or indirect determination of the biological markers of DNA damage and the changes of physical and chemical properties caused by DNA damage, several consequences of DNA damage have been identified. However, because of the complexity of molecular structure, physiological environment, the damage processes of DNA and the exact damage mechanisms of DNA are yet to be known properly.

REFERENCES

Ang, J., Song, L.Y., D'Souza, S.. et al. 2016. Mutagen synergy: Hypermutability generated by specific pairs of base analogs. *Journal of Bacteriology* 198(20); 2776–2783.

Brown, T.A. 2002. *Genomes.* 2nd edition. Oxford: Wiley-Liss; Chapter 14, Mutation, Repair and Recombination. https://www.ncbi.nlm.nih.gov/books/NBK21114/

Cannan, W.J., Pederson, D.S. 2016. Mechanisms and consequences of double-strand DNA break formation in chromatin. *Journal of Cellular Physiology* 231(1); 3–14.

Chargaff, E. 1950. Chemical specificity of nucleic acids and mechanism of their enzymatic degradation. *Experientia* 6(6); 201–209.

Chatterjee, N., Walker, G.C. 2017. Mechanisms of DNA damage, repair, and mutagenesis. *Environmental and Molecular Mutagenesis* 58(5); 235–263.

Clancy, S., 2008. *DNA Damage & Repair: Mechanisms for Maintaining DNA Integrity.* Nature Education, Cambridge, USA, 1, 103.

Colvin, O.M., Friedman, H.S., 2005. Alkylating agents. In: *Cancer: Principles and Practice of Oncology,* DeVita, V.T., Hellman, S., Rosenberg, S.A. (eds.), Lippincott Williams & Wilkins, New York, 332–344.

Deans, A.J., West, S.C. 2011. DNA interstrand crosslink repair and cancer. *Nature Review Cancer* 11(7); 467–480.

Dianfeng, L., Sansan, Y. et al. 2006. DNA damage checkpoint, damage repair, and genome stability. *Journal of Genetics and Genomics.* 5; 3–12.

Gros, L., Saparbaev, M., Laval, J. 2002. Enzymology of the repair of free radicals-induced DNA damage. *Oncogene* 21(58); 8905–8925.

Hoeijmakers, J.H. 2001. Genome maintenance mechanisms for preventing cancer. *Nature* 411(6835); 366–374.

Jun, S. 2010. DNA chemical damage and its detected. *International Journal of Chemistry* 2; 261–265.

Kino et al. 2017. Generation, repair and replication of guanine oxidation products. *Genes and Environment* 39; 1–8.

Kondo, N., Takahashi, A., Ono, K., Ohnishi, T. 2010. DNA damage induced by alkylating agents and repair pathways. *Journal of Nucleic Acids*. 2010; 1–7.

Lodish, H., et al, 2004. *Molecular Biology of the Cell*, 5th ed. Freeman, New York.

Mukherjee, A., Sasikala, W.D. 2013. Drug-DNA intercalation: From discovery to the molecular mechanism. *Advances in Protein Chemistry and Structural* 92; 1–62.

Phaniendra, A., Jestadi, D.B., Periyasamy, L. 2015. Free radicals: Properties, sources, targets, and their implication in various diseases. *Indian Journal of Clinical Biochemistry* 30(1); 11–26.

Pizzino et al. 2017. Oxidative stress: Harms and benefits for human health. *Oxidative Medicine and Cellular Longevity* 2017; 1–13.

Ramasamy, K., Shanmugam, M., Balupillai, A. *et al.* 2017. Ultraviolet radiation-induced carcinogenesis: Mechanisms and experimental models. *Journal of Radiation Cancer Research* 8(1); 4–19.

Rastogi, R.P., Richa, K.A., Tyagi, M.B., Sinha, R.P. 2010. Molecular mechanisms of ultraviolet radiation-induced DNA damage and repair. *Journal of Nucleic Acids* 2010; 1–32.

Watson, J., Crick, F. 1953. Molecular structure of nucleic acids: A structure for deoxyribose nucleic acid. *Nature* 171(4356); 737–738.

Index